Elements of Petroleum Geology

Elements of Petroleum Geology

Editor

Nakul Parashar

scitus
academics

Elements of Petroleum Geology

Edited by **Nakul Parashar**

Printed in 2017

ISBN: 978-1-68117-359-7

Library of Congress Control Number: 2015941551

SCITUS Academics LLC,
616, Corporate Way, Suite 2, 4766,
Valley Cottage, NY 10989

www.scitusacademics.com

Contents

Preface

Elements of Petroleum Geology is completely updated and revised to reflect the vast changes in the field since publication of the Second Edition. This book is a usefulprimer for geophysicists, geologists, and petroleum engineers in the oil industry who wish to expand their knowledge beyond their specialized area. It is also an excellent introductory text for a university course in petroleum geoscience. Elements of Petroleum Geology begins with an account of the physical and chemical properties of petroleum, reviewing methods of petroleum exploration and production. These methods include drilling, geophysical exploration techniques, wireline logging, and subsurface geological mapping. After describing the temperatures and pressures of the subsurface environment and the hydrodynamics of connate fluids, Selley examines the generation and migration of petroleum, reservoir rocks and trapping mechanisms, and the habit of petroleum in sedimentary basins.

Editor

Chapter 1

Synthesis and Evaluation of a Water-Soluble Hyperbranched Polymer as Enhanced Oil Recovery Chemical

Nanjun Lai[1, 2], Xiaoping Qin[1, 3], Zhongbin Ye[1, 2], Qin Peng[2], Yan Zhang[2], and Zheng Ming[2]

[1]State Key Laboratory of Oil and Gas Reservoir Geology and Exploitation, Southwest Petroleum University, Chengdu 610500, China

[2]College of Chemistry and Chemical Engineering, Southwest Petroleum University, Chengdu 610500, China

[3]College of Petroleum Engineering, Southwest Petroleum University, Chengdu 610500, China

ABSTRACT

A novel hyperbranched polymer was synthesized using acrylamide (AM), acrylic acid (AA), N-vinyl-2-pyrrolidone (NVP), and dendrite functional monomer as raw materials by redox initiation system in an aqueous medium. The hyperbranched polymer was characterized by infrared (IR) spectroscopy, ^{1}H NMR spectroscopy, ^{13}C NMR spectroscopy, elemental analysis, and scanning electron microscope (SEM). The viscosity retention rate of the hyperbranched polymer was 22.89% higher than that of the AM/AA copolymer (HPAM) at 95°C, and the viscosity retention rate was 8.17%, 12.49%, and 13.68% higher than that of HPAM in 18000 mg/L NaCl, 1800 mg/L $CaCl_2$, and 1800 mg/L $MgCl_2 \cdot 6H_2O$ brine, respectively. The hyperbranched polymer exhibited higher apparent viscosity (25.2 mPa·s versus 8.1 mPa·s) under 500 s^{-1} shear rate at 80°C. Furthermore, the enhanced oil recovery (EOR) of 1500 mg/L hyperbranched polymer solutions was up to 23.51% by the core flooding test at 80°C.

INTRODUCTION

Polymer flooding plays an important role in enhanced oil recovery (EOR) [1–7]. However, the most widely used water-soluble polymers, polyacrylamide and partially hydrolyzed polyacrylamide, are not suitable for high temperature, high salinity, and high flow rate injection owing to hydrolysis, decomposition, degradation, shear damage, and so forth [1, 6, 8–11]. With the growing demand for petroleum resources, the water-soluble polymer, which displays perfect temperature-resistance, salt-resistance, and shear-resistance in harsh conditions, is a challenge to the oil filed chemists [12].

In recent decades, many studies demonstrated that acrylamide (AM) copolymerized with an applicable functional monomer, such as N,N-dimethylacrylamide, methacrylamide, N-vinyl-2-pyrrolidone (NVP), 2-acrylamido-2-methyl-1-propanesulfonic acid (AMPS), sodium allylsulfonate, acrylic acid (AA), and ethylenesulfonic acid, could obtain more satisfying polymer possessing better temperature-resistance and salt-resistance for EOR [12–17]. What is more, dendrite polymers, star polymers, and hyperbranched polymers, which may exhibit excellent shear-resistance performance due to their special network structure, have been reported widely in many other applied fields [18–25]. This special network structure may reduce the effect of shear on polymer molecular chain, which may recover to a certain extent after being cut and eventually obtaining higher viscosity retention rate.

Keeping in mind all the above points, herein, a novel hyperbranched polymer was synthesized by free radical polymerization based on AM, AA, NVP, and dendrite functional monomer aiming to obtain satisfying temperature-resistance, salt-resistance, and shear-resistance.

EXPERIMENTAL

Chemicals and Reagents

Ethylenediamine (EDA, AR), methyl acrylate (AR), methanol (AR), ethanol (AR), N,N-dimethylformamide (DMF, AR), maleic anhydride (AR), acrylic acid (AR), acrylamide (AR), N-vinyl-2-pyrrolidone (NVP, AR), sodium hydroxide (AR), sodium hydrogen sulfite ($NaHSO_3$, AR), ammonium persulfate ($(NH_4)_2S_2O_8$, AR), sodium chloride (NaCl, AR), calcium chloride anhydrous ($CaCl_2$, AR), magnesium chloride hexahydrate ($MgCl_2 \cdot 6H_2O$, AR), potassium chloride (KCl, AR), sodium sulfate (Na_2SO_4, AR), and sodium bicarbonate ($NaHCO_3$, AR) were purchased from Chengdu Kelong Chemical Reagent Factory (Sichuan, China). All chemicals and reagents were used as received without any further purification.

Synthesis of Dendrite Functional Monomer

Synthesis of 0.5 generation dendritic macromolecule ($DM_{0.5}$): 113.5 g methyl acrylate was added into a three-necked flask with methanol as solvent, and then 9.0 g ethylenediamine was dripped into the stirred solution in the three-necked flask. The reaction time was 24 h at 25°C. After reaction, the product was purified by vacuum distillation and silica gel column. Then the $DM_{0.5}$ was obtained [18, 20].

Synthesis of 1.0 generation dendritic macromolecule ($DM_{1.0}$): 40.0 g ethylenediamine was added into a three-necked flask with methanol as solvent, then 20.2 g $DM_{0.5}$ was dripped into the stirred solution in the three-necked flask. The reaction time was 48 h at 25°C, and the product was the $DM_{1.0}$, which was purified by vacuum distillation and silica gel column [18, 20].

Modification of $DM_{1.0}$: 4.4 g maleic anhydride was added into a round-bottom flask with N,N-dimethylformamide as solvent, and 8.0 g $DM_{1.0}$ was dripped into the round-bottom flask. The reaction time was 8 h at 70°C, and the dendrite functional monomer was obtained by vacuum filtration.

Synthesis of Hyperbranched Polymer and HPAM

Firstly, 7.00 g AM, 2.95 g AA, 0.01 g dendrite functional monomer, 0.04 g NVP, and 1.65 g sodium hydroxide were added into a 100 mL three-necked flask with 38.35 mL distilled water as solvent. Secondly, 0.04 g $NaHSO_3$-$(NH_4)_2S_2O_8$ initiator (mol ratio = 1 : 1) was taken along with distilled water in the three-necked flask. And then, the copolymerization was carried out for 4 h at 50°C under nitrogen atmosphere. Finally, the hyperbranched polymer was obtained by ethanol washing, drying, and pulverizing. The synthesis route of the hyperbranched polymer is shown in Scheme 1.

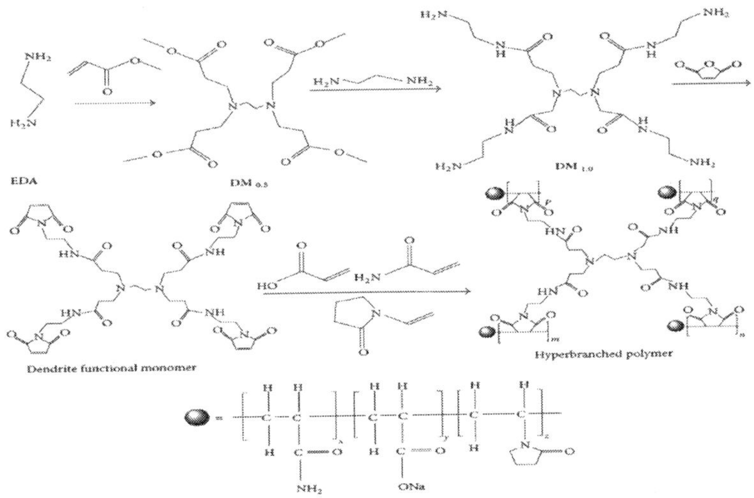

Scheme 1: The synthesis route of the hyperbranched polymer.

The AM/AA copolymer (HPAM) was synthesized using 7.00 g AM and 2.95 g AA as raw materials through the same synthesis method.

Characterization

Infrared (IR) spectra of the dendrite functional monomer and hyperbranched polymer were measured with KBr pellets using Perkin Elmer RX-1 spectrophotometer (Beijing Rayleigh Analytical Instrument, China). ^{1}H NMR spectroscopy and ^{13}C NMR spectroscopy of the dendrite functional monomer, hyperbranched polymer, and HPAM were recorded on Bruker AC-E 200 spectrometer (Bruker BioSpin, Switzerland) at 400 MHz with D_2O as solvent. The elemental analysis of the hyperbranched polymer and HPAM was carried out through Vario EL III elemental analyzer (Wuxi Chuangxiang Analytical Instrument, China). The microstructures of the hyperbranched polymer and HPAM were observed via Quanta 450 scanning electron microscope (SEM, FEI Company, USA).

Weight-Average Molecular Weight

The weight-average molecular weight (M_w) of the hyperbranched polymer and HPAM was determined by a BI-200SM wide angle dynamic/static laser light scattering apparatus at 25°C. The laser wavelength was 532 nm. The M_w of the hyperbranched polymer and HPAM can be obtained by the following equation [26, 27]:

$$\frac{KC}{R_{vv}(q)} \cong \frac{1}{M_w}\left(1 + \frac{1}{3}\langle R_g \rangle^2 q^2\right), \tag{1}$$

where C is the concentration of polymer solution, g/mL; K is a constant; $\langle R_g \rangle$ is the average radius of gyration, nm; $R_{VV}(q)$ is the Rayleigh ratio; and $q = (4\pi n/\lambda_o)\sin(\theta/2)$ with n, λ_o, and θ being the solvent refractive index, the wavelength of laser in vacuo, and the scattering angle, respectively.

Temperature-Resistance and Salt-Resistance

Hyperbranched polymer and HPAM solutions (5000 mg/L) were prepared with distilled water. The apparent viscosity of these polymers solutions was tested using Brookfiled DV-III viscometer at different temperatures. The salt-resistance performance was studied by increasing salt (NaCl, $CaCl_2$, or $MgCl_2 \cdot 6H_2O$) concentration, and then the apparent viscosity of these polymers solutions was measured via Brookfield DV-III viscometer at 20°C.

Shear-Resistance

Shear-resistance of the hyperbranched polymer and HPAM solutions (5000 mg/L) was measured using HAAKE RS 6000 rotational rheometer (Thermo Fisher Scientific, Germany) at 80°C [12, 15–17]. The samples were prepared with distilled water.

Core Flooding Experiments

Two Berea sandstone cores were used to study the EOR ability of these copolymer solutions (1500 mg/L) prepared with brine. Total dissolved solids (TDS) and chemical composition of the brine are listed in Table 1. The core was placed into Hassler core holder with 1.0 MPa backpressure and 3.0 MPa confining pressure. It was saturated with the brine, and then it was saturated with crude oil (62.2 mPa·s at 80°C) at different injection rate (0.1-0.2 mL/min) until irreducible water saturation (S_{wi}) was established. After 96 h of aging, the brine was injected at 0.2 mL/min to displace the crude oil until water cut reached 95%, and then the polymer solution was injected at 0.2 mL/min to obtain water cut 95% once more. The EOR of polymer solutions is calculated with the following equation:

$$\mathrm{EOR} = E - E_w, \tag{2}$$

where EOR is enhanced oil recovery of polymer solution, %; E is the oil recovery of water flooding and polymer flooding process, %; E_w is the oil recovery of water flooding process, %.

Table 1: TDS and chemical composition of the brine

Composition	NaCl	KCl	$CaCl_2$	$MgCl_2·6H_2O$	Na_2SO_4	$NaHCO_3$	TDS
Content (g/L)	8.007	0.155	0.762	2.183	0.138	0.516	10.599

All core flooding experiments were conducted at 80°C. The maximum work pressure of the ISCO pump is 50 MPa, and its maximum and minimum displacement rates are 50.000 and 0.001 mL/min, respectively. The pressure drop was recorded by a pressure sensor with a precision of ±0.0001 MPa. And flow chart of the core flooding tests is shown in Figure 1.

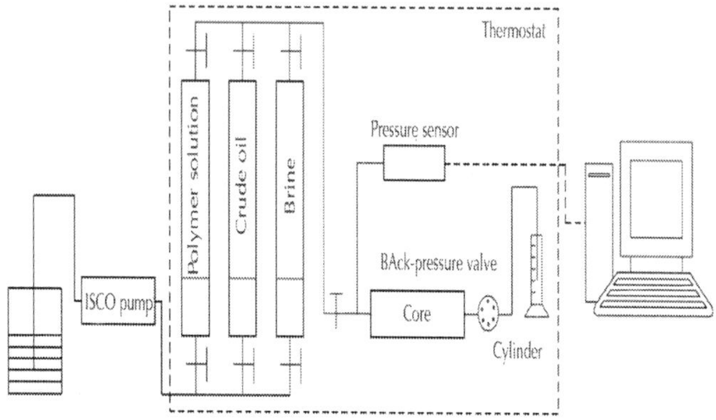

Figure 1: Flow chart of the core flooding tests.

RESULTS AND DISCUSSION

IR Spectra Analysis

The structures of the dendrite functional monomer and hyperbranched polymer were confirmed by IR spectra as illustrated in Figure 2. The dendrite functional monomer, which was prepared using EDA, methyl acrylate, and maleic anhydride, was confirmed by strong absorptions at 3383.78 cm^{-1} (–NH stretching vibration), 2949.63 cm^{-1} ($-CH_2$ stretching vibration), 1646.91 cm^{-1} (C=O stretching vibration and carbon double-bond stretching vibration), and 1554.98 cm^{-1} (C–N stretching vibration and –NH bending vibration) in the IR spectroscopy of the dendrite functional monomer. Pure NVP exhibited a very strong absorption at 1703.01 cm^{-1}, which reflected the carbonyl stretching; at 1629.32 cm^{-1}, which was a carbon double-bond stretching vibration; and at 1445.11 cm^{-1}, which was the characteristic absorption peak of NVP. The characteristic absorptions of the dendrite functional monomer and NVP were clearly presented, and the carbon double-bond was not detected in the IR spectroscopy of the hyperbranched polymer.

As expected, the IR spectra demonstrated that the hyperbranched polymer was successfully synthesized.

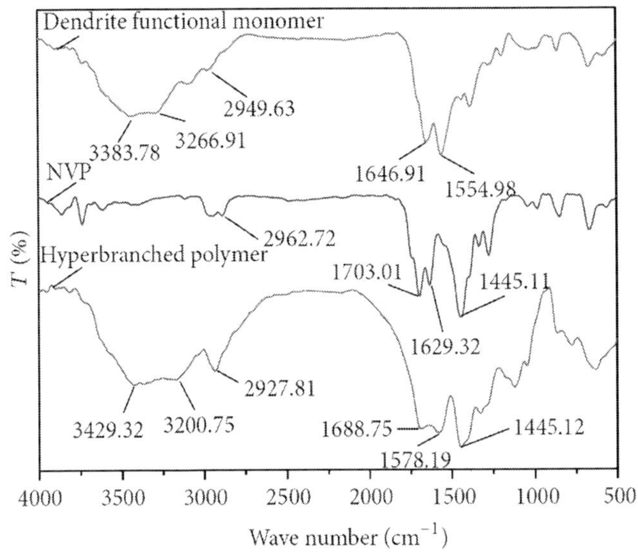

Figure 2: IR spectra of the dendrite functional monomer, NVP, and hyperbranched polymer.

^{1}H NMR and ^{13}C NMR Analyses

The ^{1}H NMR spectrum and ^{13}C NMR spectrum of the dendrite functional monomer are shown in Figures 3(a) and 3(b), respectively. In Figure 3(a), the chemical shift value at 2.44 ppm is assigned to the –CH_2–CH_2–C (O)–NH– protons. The chemical shift value at 2.60 ppm is due to the –CH_2–N (CH_2–CH_2–C (O)–)$_2$ protons. The –CH_2–CH_2–C (O)–NH– protons appear at 2.79 ppm. The chemical shift value at 3.04 ppm is assigned to the –NH–CH_2–CH_2– protons. The –NH–CH_2–CH_2– protons appear at 3.42 ppm. The chemical shift value at 5.95 ppm is due to the –C (O)–CH=CH–C(O)– protons. In Figure 3(b), the chemical shift value at 171.11 ppm is due to C_6. The chemical shift value at 164.96 ppm belongs to C_5. The chemical shift value at 134.49 ppm is due to C_4. The characteristic peak at

48.88 ppm belongs to C_3. The characteristic peak at 36.90 ppm is due to C_2. The chemical shift value at 31.01 ppm is assigned to C_1. The results of ^1H NMR spectrum and ^{13}C NMR spectrum showed that the dendrite functional monomer was synthesized.

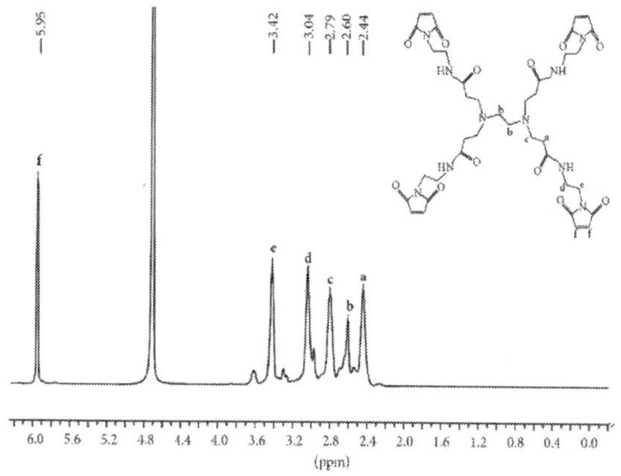

(a)

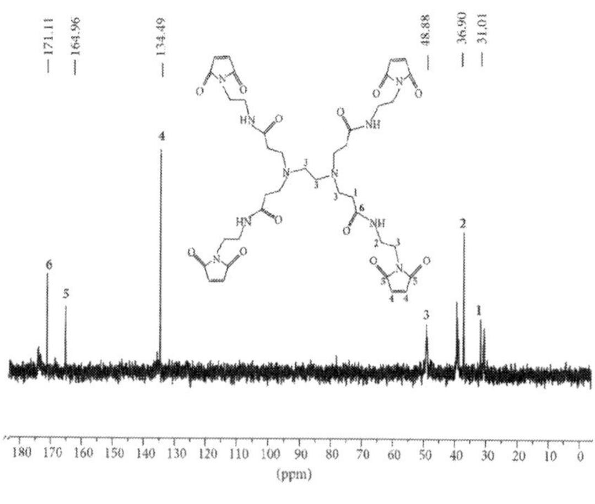

(b)

Figure 3: (a) ^1H NMR spectrum of the dendrite functional monomer in D_2O; (b) ^{13}C NMR spectrum of the dendrite functional monomer in D_2O.

The ^1H NMR spectrum and ^{13}C NMR spectrum of the hyperbranched polymer are shown in Figures 4(a) and 4(b), respectively. In Figure 4(a), the chemical shift value at 3.23 ppm is due to the $-NH-CH_2-CH_2-$ protons and the $-C\ (O)-CH_2-CH_2-CH_2-$ protons. The chemical shift value at 2.65 ppm is assigned to the $-CH_2-N\ (CH_2-CH_2-C\ (O)-)_2$ protons. The $-CH_2-CH_2-C\ (O)-NH-$ protons and the $-CH\ (C\ (O)-)-CH(C(O)-)-$ protons appear at 2.49 ppm. The chemical shift value at 2.16 ppm is assigned to the $-C(O)-CH_2-CH_2-CH_2-$ protons, the $-CH_2-CH_2-C(O)-NH-$ protons, and the $-CH_2-$ protons which are obtained from the carbon double-bonds of AM, AA, and NVP. The –CH– protons, which are free radical polymerization products of the carbon double-bonds of AM, AA, and NVP, appear at 1.54 ppm. In Figure 4(b), the chemical shift value at 182.96 ppm belongs to C_8. The chemical shift value at 179.79 ppm is due to C_7. The chemical shift value at 178.63 ppm is assigned to C_6. The chemical shift value at 44.80 ppm is due to C_5. The characteristic peak at 42.30 ppm belongs to C_4. The characteristic peak from 35.11 to 36.95 ppm is due to C_3. The chemical shift value at 31.80 ppm is assigned to C_2. And the characteristic peak of C_1 is observed at 17.62 ppm. ^1H NMR spectrum and ^{13}C NMR spectrum indicated that the hyperbranched polymer was successfully synthesized.

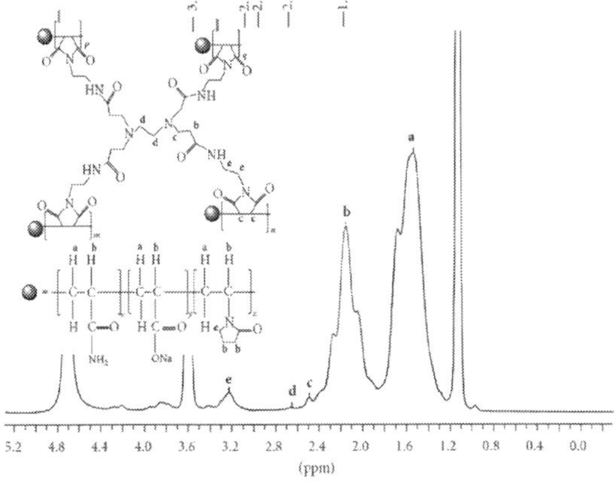

(a)

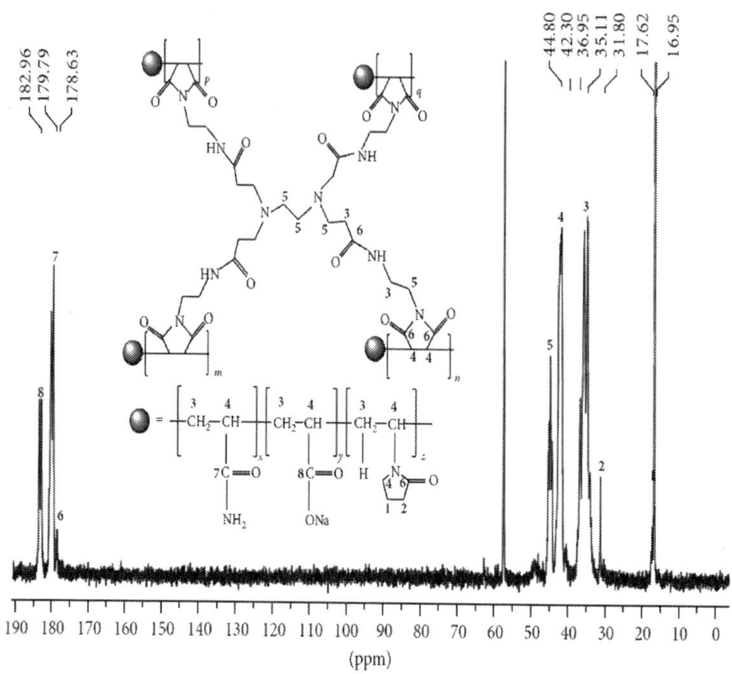

(b)

Figure 4: (a) ^1H NMR spectrum of the hyperbranched polymer in D_2O; (b) ^{13}C NMR spectrum of the hyperbranched polymer in D_2O.

The ^1H NMR spectrum and ^{13}C NMR spectrum of HPAM are shown in Figures 5(a) and 5(b), respectively. In Figure 5(a), the chemical shift value at 2.12 ppm is assigned to the –CH–CH_2– protons. The characteristic peak of the –CH–CH_2– protons appears at 1.55 ppm. In Figure 5(b), the chemical shift value at 183.12 ppm belongs to C_4. The chemical shift value at 179.58 ppm is due to C_3. The characteristic peak at 41.95 ppm belongs to C_2. The chemical shift value at 34.99 ppm is assigned to C_1.

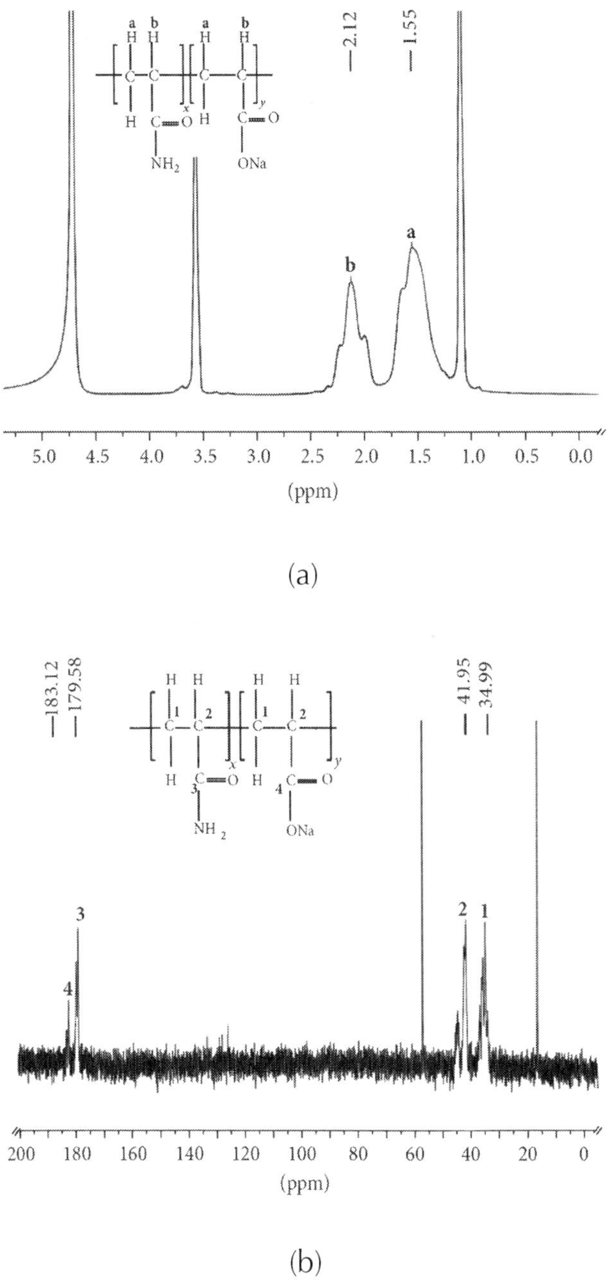

Figure 5: (a) ^{1}H NMR spectrum of HPAM in D_2O; (b) ^{13}C NMR spectrum of HPAM in D_2O.

Elemental Analysis of the Hyperbranched Polymer and HPAM

The elemental analysis of the hyperbranched polymer and HPAM was carried out by Vario EL III elemental analyzer. The content of different elements can be calculated by detecting the gases, which are the decomposition products of these copolymers at high temperature. Theoretical values of the hyperbranched polymer are 50.49% (C %), 6.53% (H%), 30.12% (O%), and 12.86% (N%); found values of the hyperbranched polymer are 45.46% (C%), 6.12% (H%), 26.73% (O%), and 11.36% (N%). Theoretical values of HPAM are 50.50% (C %), 6.60% (H %), 29.03% (O%), and 13.87% (N%); found values of HPAM are 45.91% (C%), 6.04% (H%), 26.19% (O%), and 12.03% (N%).

Microscopic Structure Analysis by SEM

The microscopic structures of the HPAM and hyperbranched polymer solutions (2000 mg/L) prepared with distilled water were observed through SEM at room temperature. Among these images, Figures 6(a)–6(c) are HPAM solutions at different scan sizes (20 μm, 5000x; 10 μm, 10000x; 5 μm, 20000x, resp.). Similarly, the images of hyperbranched polymer solutions are shown in Figure 6(d) (20 μm, 5000x), Figure 6(e) (10 μm, 10000x), and Figure 6(f) (5 μm, 20000x). As shown in Figure 6, it could be obviously observed that there were space net structures in the images of the hyperbranched polymer solutions. Moreover, it could be found that the microscopic reticular structures of the hyperbranched polymer solutions were much more compact than those of HPAM solutions in the same scan size. The much denser networks of the hyperbranched polymer solutions may help to reduce the effect of shear on the hyperbranched polymer molecular chain and improve the viscosity retention rate of the hyperbranched polymer at high shear rate.

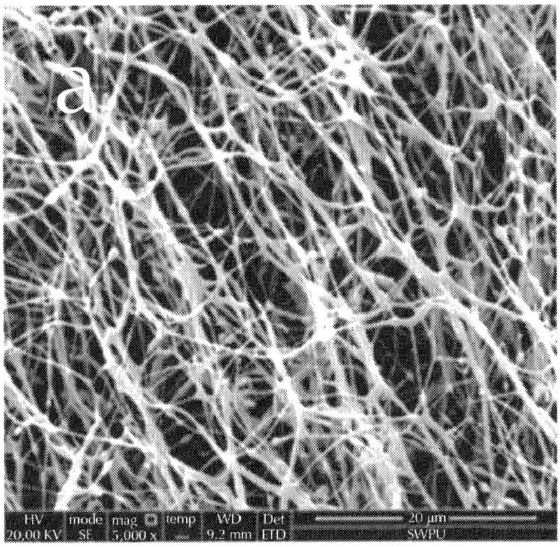
a
HV mode mag temp WD Det
20.00 KV SE 5,000 x 9.2 mm ETD
20 µm
SWPU

(a)

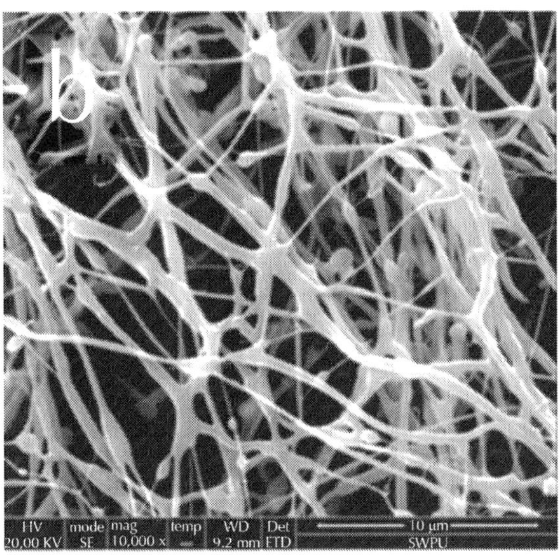
b
HV mode mag temp WD Det
20.00 KV SE 10,000 x 9.2 mm ETD
10 µm
SWPU

HPAM

(b)

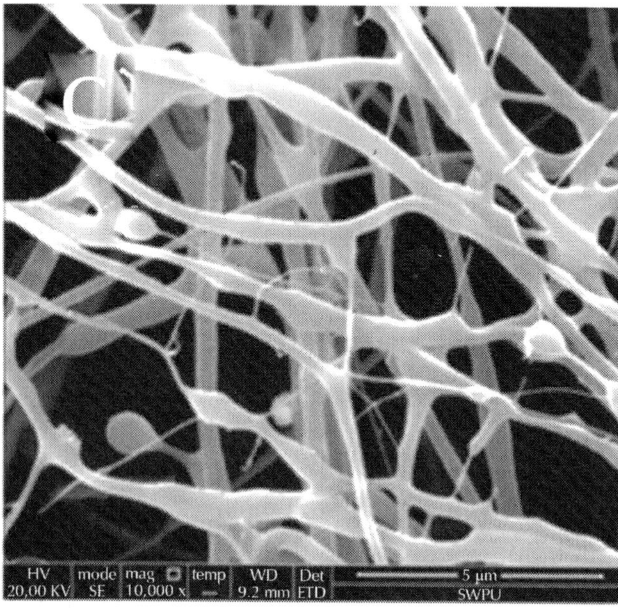
c
HV
20,00 KV
mode
SE
mag
10,000 x
temp
WD
9.2 mm
Det
ETD
5 µm
SWPU

(c)

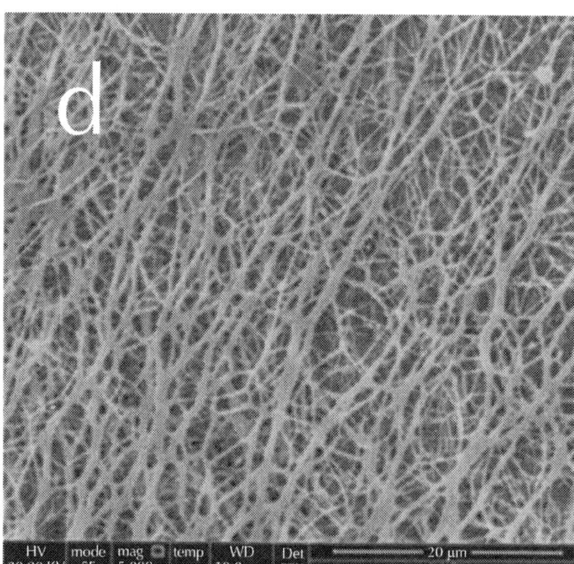
d
HV
20,00 KV
mode
SE
mag
5,000 x
temp
WD
10.0 mm
Det
ETD
20 µm
SWPU

(d)

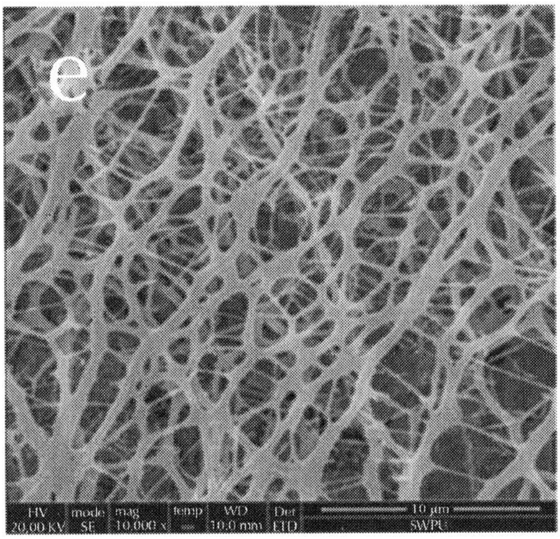

Hyperbranched polymer

(e)

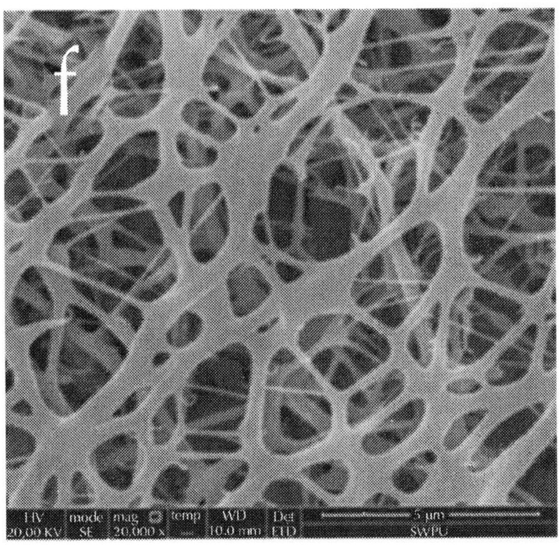

(f)

Figure 6: SEM images of HPAM and hyperbranched polymer: (a) HPAM solution at 20 μm, 5000x; (b) HPAM solution at 10 μm, 10000x; (c) HPAM solution at 5 μm, 20000x; (d) hyperbranched polymer solution at 20 μm, 5000x; (e) hyperbranched polymer solution at 10 μm, 10000x; (f) hyperbranched polymer solution at 5 μm, 20000x.

Weight-Average Molecular Weight

2 mg/L copolymer solution was prepared using distilled water and filtered by a 0.5 μm Millipore Millex-LCR filter. As shown in Figure 7, the weight -average molecular weight of the hyperbranched polymer and HPAM was 5.72 and 5.64 × 10^6 g/mol, respectively. The hyperbranched polymer has higher weight-average molecular weight than HPAM due to its hyperbranched structure.

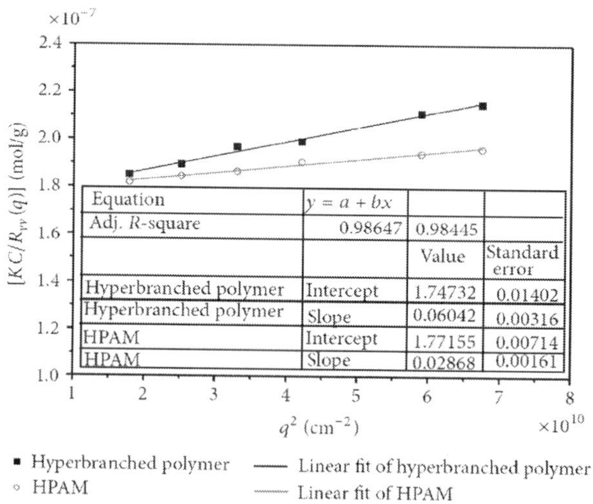

Figure 7: KC / R_{vv}(q) versus q^2 for the hyperbranched polymer and HPAM.

Temperature-Resistance

The apparent viscosity versus temperature curves of the hyperbranched polymer and HPAM solutions is shown in Figure 8. Compared with HPAM, the hyperbranched polymer displayed better temperature-resistance (apparent viscosity: 558.6 mPa·s versus 260.3 mPa·s and viscosity retention rate: 45.01% versus 22.12%) at 95°C. This phenomenon may be explained by the inelasticity structure of pyrrole ring which can improve the thermal stability of the hyperbranched polymer.

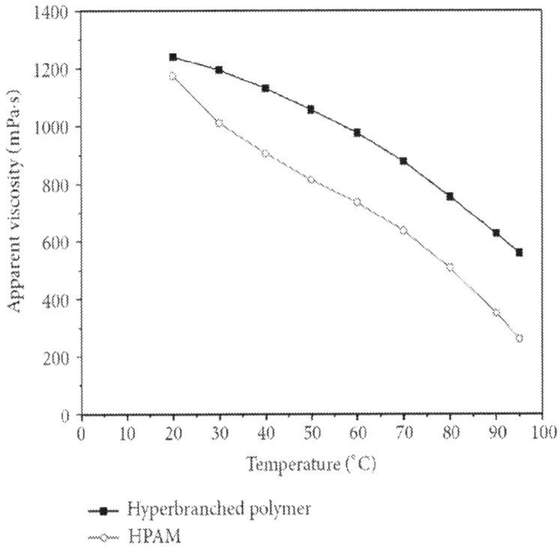

Figure 8: Apparent viscosity versus temperature for HPAM and hyperbranched polymer solutions (5000 mg/L).

Salt-Resistance

The influences of salt (NaCl, $CaCl_2$, or $MgCl_2 \cdot 6H_2O$) on apparent viscosity of the HPAM and hyperbranched polymer solutions were carried out at 20°C. As shown in Figures 9(a)–9(c), with the increase of salt concentration (NaCl, $CaCl_2$, or $MgCl_2 \cdot 6H_2O$), the apparent viscosity of HPAM solutions decreased rapidly, and then it is kept at a low value. Similarly, the measure of the hyperbranched polymer solutions displayed similar phenomena. However, compared with HPAM solution, the hyperbranched polymer solutions displayed better antisalt due to higher apparent viscosity under the same conditions. These results reveal that the hyperbranched polymer can withstand higher salt concentration than HPAM. This characteristic may be well explained by the special network structure which can enhance the interaction between the hyperbranched polymer chains, and crimping degree of the polymeric chains will be smaller than HPAM at the same salt concentration. Thus the hyperbranched polymer exhibits higher apparent viscosity and retention rates.

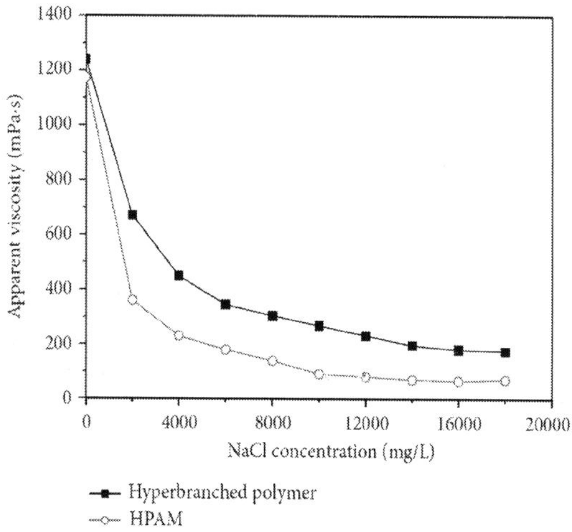

(a)

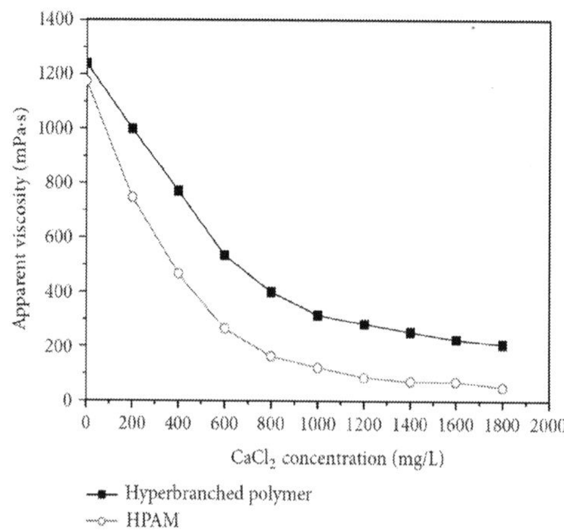

(b)

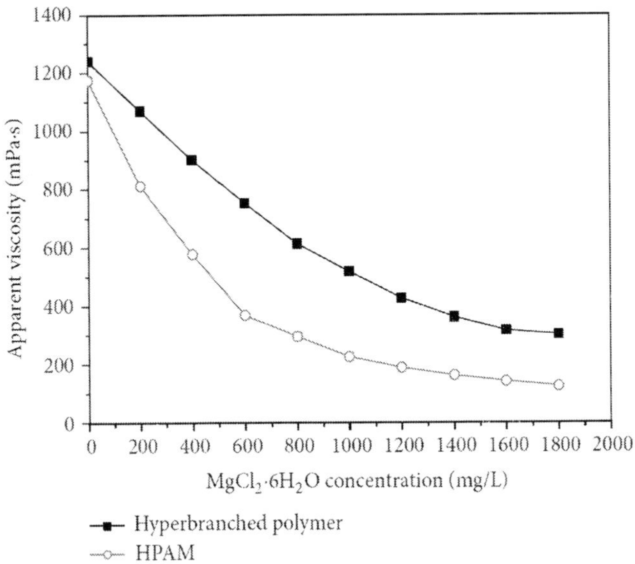

(c)

Figure 9: Salt-resistance ((a) NaCl, (b) $CaCl_2$, and (c) $MgCl_2 \cdot 6H_2O$) of HPAM and hyperbranched polymer solutions (5000 mg/L) at 20°C.

Shear-Resistance

Shear-resistance of the polymer solutions was conducted on HAAKE RS 6000 rotational rheometer at 80°C by changing the shear rate from 170 s^{-1} to 500 s^{-1} and from 500 s^{-1} to 170 s^{-1} around. As shown in Figure 10, the viscosity retention rate of the HPAM and the hyperbranched polymer was 61.95% and 91.64%, respectively, when one cycle was completed. The phenomena may support the microscopic reticular structures of the hyperbranched polymer which can reduce the effect of shear on the hyperbranched polymer molecular chain during shear process and restore the structures of the hyperbranched polymer after being sheared.

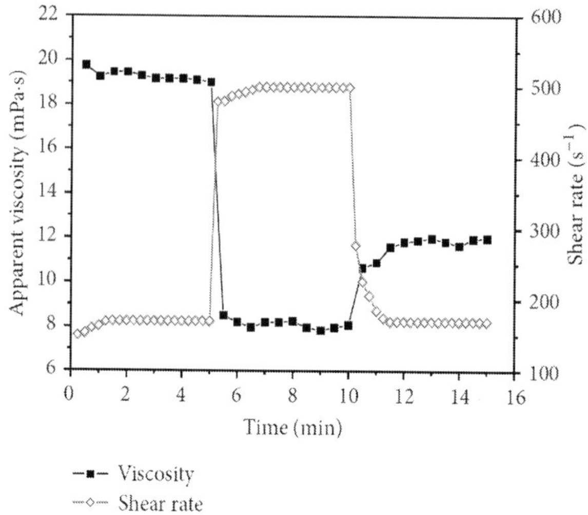

(a)

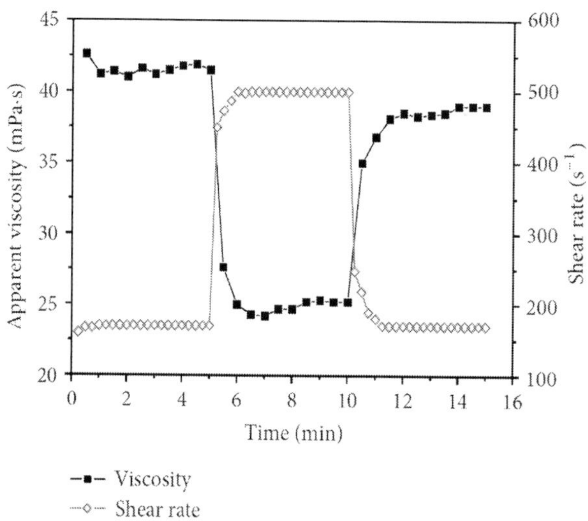

(b)

Figure 10: Shear-resistance of HPAM (a) and hyperbranched polymer (b) at 80°C (5000 mg/L).

Enhanced Oil Recovery

As shown in Table 2, the EOR of the hyperbranched polymer solutions and HPAM solutions was 23.51%, and 16.67%, respectively. This phenomenon may support higher viscosity retention rate of the hyperbranched polymer contributes to expand injection water sweeping volume and enhance oil recovery. As shown in Figure11, compared with HPAM, the hyperbranched polymer revealed stronger ability of establishing flow resistance and reducing water cut in polymer flooding. This phenomenon may support the sweep efficiency which is obviously improved by the hyperbranched polymer due to its excellent temperature-resistance, salt-resistance, and shear-resistance.

Table 2: The parameters of cores and the results of core flooding tests

Polymer	Cores	Length (cm)	Diameter (cm)	Porosity (%)	Permeability (mD)	S_{wi}(%)	E(%)	E_w(%)	EOR (%)
HPAM	1#	8.95	3.79	23.75	985.28	20.42	47.63	30.97	16.67
Hyperbra-nched polymer	2#	8.87	3.78	23.69	980.15	20.16	54.33	30.82	23.51

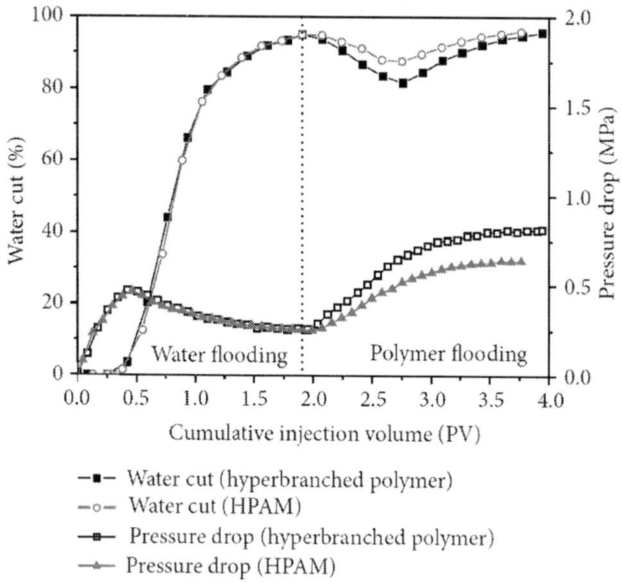

Figure 11: EOR results of the HPAM and hyperbranched polymer solutions at 80°C.

CONCLUSIONS

A novel hyperbranched polymer possessing microscopic reticular structure was successfully synthesized using AM, AA, NVP, and dendrite functional monomer as raw materials under mild conditions. Compared with HPAM, the hyperbranched polymer exhibits obvious advantages in temperature-resistance, salt-resistance, and shear-resistance due to the introduction of pyrrole ring which can reduce the influence of high temperature on the polymer molecular chain and the introduction of the reticular structures which can be favorable to decrease the crimping degree of polymeric chains under high shear rate and high salinity. Thus, the EOR capability of the hyperbranched polymer is improved remarkably even in a harsh condition.

ACKNOWLEDGMENTS

This work was financially supported by the Open Fund (PLN1212) of State Key Laboratory of Oil and Gas Reservoir Geology and Exploitation (Southwest Petroleum University), the Specialized Research Fund for the Doctoral Program of Higher Education (20125121120011), the Key Program for Undergraduate Extracurricular Open Experiment (KSZ1246/KSZ1247) of Southwest Petroleum University, and the Cultivation Project of Sichuan Province Science and Technology Innovation Seedling Project (20132060).

REFERENCES

1. N. Mungan, F. W. Smith, J. L. Thompson, O. Sinclair, and C. Gas, "Some aspects of polymer floods,"Journal of Petroleum Technology, vol. 18, no. 9, pp. 1143–1150, 1966.
2. G. Li, L. Zhai, G. Xu, Q. Shen, H. Mao, and M. Pei, "Current Tertiary oil recovery in China," Journal of Dispersion Science and Technology, vol. 21, no. 4, pp. 367–408, 2000.
3. T. Masuda, S.-I. Yamamoto, O. Moriya, M. Kashio, and T. Sugizaki, "Preparation of stimuli-responsive polysilsesquioxane grafted block copolymer of acrylamide monomers," Polymer Journal, vol. 40, no. 2, pp. 126–136, 2008.
4. B. Yadali Jamaloei, R. Kharrat, and F. Torabi, "Analysis and correlations of viscous fingering in low-tension polymer flooding in heavy oil reservoirs," Energy and Fuels, vol. 24, no. 12, pp. 6384–6392, 2010.
5. L. Shi, Z. Ye, Z. Zhang, C. Zhou, S. Zhu, and Z. Guo, "Necessity and feasibility of improving the residual resistance factor of polymer flooding in heavy oil reservoirs," Petroleum Science, vol. 7, no. 2, pp. 251–256, 2010.

6. D. A. Z. Wever, F. Picchioni, and A. A. Broekhuis, "Polymers for enhanced oil recovery: a paradigm for structure-property relationship in aqueous solution," Progress in Polymer Science, vol. 36, no. 11, pp. 1558–1628, 2011.
7. Z. Q. Guo, M. Z. Dong, Z. X. Chen, and J. Yao, "Dominant scaling groups of polymer flooding for enhanced heavy oil recovery," Industrial & Engineering Chemistry Research, vol. 52, no. 2, pp. 911–921, 2013.
8. S. H. Chang and I. J. Chung, "Effect of shear flow on polymer desorption and latex dispersion stability in the presence of adsorbed polymer," Macromolecules, vol. 24, no. 2, pp. 567–571, 1991.
9. T. Rho, J. Park, C. Kim, H.-K. Yoon, and H.-S. Suh, "Degradation of polyacrylamide in dilute solution,"Polymer Degradation and Stability, vol. 51, no. 3, pp. 287–293, 1996.
10. L. Xue, U. S. Agarwal, and P. J. Lemstra, "Shear degradation resistance of star polymers during elongational flow," Macromolecules, vol. 38, no. 21, pp. 8825–8832, 2005.
11. R. Zhang, Z. Ye, L. Peng, N. Qin, Z. Shu, and P. Luo, "The shearing effect on hydrophobically associative water-soluble polymer and partially hydrolyzed polyacrylamide passing through wellbore simulation device," Journal of Applied Polymer Science, vol. 127, no. 1, pp. 682–689, 2013.
12. Z. B. Ye, G. J. Gou, S. H. Gou, W. C. Jiang, and T. Y. Liu, "Synthesis and characterization of a water-soluble sulfonates copolymer of acrylamide and N-Allylbenzamide as enhanced oil recovery chemical,"Journal of Applied Polymer Science, vol. 128, no. 3, pp. 2003–2011, 2013.
13. A. Sabhapondit, A. Borthakur, and I. Haque, "Water soluble acrylamidomethyl propane sulfonate (AMPS) copolymer as an enhanced oil recovery chemical," Energy and Fuels, vol. 17, no. 3, pp. 683–688, 2003.
14. C.-S. Yang, K. Shin, and H. K. Jeong, "Thermal analysis of poly(sodium 4-styrenesulfonate) intercalated graphite oxide composites," Chemical Physics Letters, vol. 517, no. 4–6, pp. 196–198, 2011.

15. X.-J. Liu, W.-C. Jiang, S.-H. Gou, Z.-B. Ye, and X.-D. Xie, "Synthesis and evaluation of a water-soluble acrylamide binary sulfonates copolymer on MMT crystalline interspace and EOR," Journal of Applied Polymer Science, vol. 125, no. 2, pp. 1252–1260, 2012.
16. X. J. Liu, W. C. Jiang, S. H. Gou, Z. B. Ye, and C. Luo, "Synthesis and clay stabilization of a water-soluble copolymer based on acrylamide, modular -cyclodextrin, and AMPS," Journal of Applied Polymer Science, vol. 128, no. 5, pp. 3398–3404, 2012.
17. X. J. Liu, W. C. Jiang, S. H. Gou et al., "Synthesis and evaluation of novel water-soluble copolymers based on acrylamide and modular -cyclodextrin," Carbohydrate Polymers, vol. 96, pp. 47–56, 2013.
18. D. A. Tomalia, H. Baker, J. Dewald et al., "New class of polymers: starburst-dendritic macromolecules,"Polymer Journal, vol. 17, no. 1, pp. 117–132, 1984.
19. D. A. Tomalia, H. Baker, J. Dewald et al., "Dendritic macromolecules: synthesis of starburst dendrimers," Macromolecules, vol. 19, no. 9, pp. 2466–2468, 1986.
20. J. C. Hummelen, J. L. J. Van Dongen, and E. W. Meijer, "Electrospray mass spectrometry of poly(propylene imine) dendrimers-The issue of dendritic purity or polydispersity," Chemistry, vol. 3, no. 9, pp. 1489–1493, 1997.
21. L. J. Twyman, A. E. Beezer, R. Esfand, M. J. Hardy, and J. C. Mitchell, "The synthesis of water soluble dendrimers, and their application as possible drug delivery systems," Tetrahedron Letters, vol. 40, no. 9, pp. 1743–1746, 1999.
22. M. S. Diallo, S. Christie, P. Swaminathan et al., "Dendritic chelating agents. 1. Cu(II) binding to ethylene diamine core poly(amidoamine) dendrimers in aqueous solutions," Langmuir, vol. 20, no. 7, pp. 2640–2651, 2004.
23. L. Xue, U. S. Agarwal, M. Zhang et al., "Synthesis and direct topology visualization of high-molecular-weight star PMMA," Macromolecules, vol. 38, no. 6, pp. 2093–2100, 2005.

24. D. S. Achilleos, T. K. Georgiou, and C. S. Patrickios, "Amphiphilic model conetworks based on cross-linked star copolymers of benzyl methacrylate and 2-(dimenthylamino) ethyl methacrylate: synthesis, characterization, and DNA adsorption studies," Biomacromolecules, vol. 7, no. 12, pp. 3396–3405, 2006.
25. X. Y. Wang, Y. C. Zhang, T. F. Li, W. D. Tian, Q. Zhang, and Y. Y. Cheng, "Generation 9 polyamidoamine dendrimer encapsulated platinum nanoparticle mimics catalase size, shape, and catalytic activity," Langmuir, vol. 29, no. 17, pp. 5262–5270, 2013.
26. X. Wang, X. Qiu, and C. Wu, "Comparison of the coil-to-globule and the globule-to-coil transitions of a single poly(N-isopropylacrylamide) homopolymer chain in water," Macromolecules, vol. 31, no. 9, pp. 2972–2976, 1998.
27. H. Chen, J. Li, Y. Ding, G. Zhang, Q. Zhang, and C. Wu, "Folding and unfolding of individual PNIPAM-g-PEO copolymer chains in dilute aqueous solutions," Macromolecules, vol. 38, no. 10, pp. 4403–4408, 2005.

Chapter 2

Synthesis and Performance of an Acrylamide Copolymer Containing Nano-SiO_2 as Enhanced Oil Recovery Chemical

Zhongbin Ye[1, 2], Xiaoping Qin[1, 2], Nanjun Lai[1, 2], Qin Peng[2] Xi Li[2], and Cuixia Li[3]

[1]State Key Laboratory of Oil and Gas Reservoir Geology and Exploitation, Southwest Petroleum University, Chengdu 610500, China

[2]College of Chemistry and Chemical Engineering, Southwest Petroleum University, Chengdu 610500, China

[3]Department of Petroleum Chemical Engineering, Karamay Vocational & Technical College, Karamay 833600, China

ABSTRACT

A novel copolymer containing nano-SiO_2 was synthesized by free radical polymerization using acrylamide (AM), acrylic acid (AA), and nano-SiO_2 functional monomer (NSFM) as raw materials under mild conditions. The AM/AA/NSFM copolymer was characterized by infrared (IR) spectroscopy, ^1H NMR spectroscopy, elemental analysis, and scanning electron microscope (SEM). It was found that the AM/AA/NSFM copolymer exhibited higher viscosity than the AM/AA copolymer at $500\,s^{-1}$ shear rate (18.6 mPa·s versus 8.7 mPa·s). It was also found that AM/AA/NSFM could achieve up to 43.7% viscosity retention rate at 95°C. Mobility control results indicated that AM/AA/NSFM could establish much higher resistance factor (RF) and residual resistance factor (RRF) than AM/AA under the same conditions (RF: 16.52 versus 12.17, RRF: 3.63 versus 2.59). At last, the enhanced oil recovery (EOR) of AM/AA/NSFM was up to 20.10% by core flooding experiments at 65°C.

INTRODUCTION

Polymer flooding plays an important role in the field of enhanced oil recovery (EOR) [1, 2]. However, the current widely used polymers, polyacrylamide (PAM) and partially hydrolyzed polyacrylamide (HPAM), cannot completely meet the requirements due to the hydrolysis, degradation, and others under high temperature or high salinity [3–6]. Furthermore, PAM and HPAM have poor shear resistance [2–7]. Polymer molecular chains will be cut off when polymer solution passes through the pump, pipeline, perforation, and porous medium at high speed, so the viscosity of polymer solution will be greatly reduced [1, 7, 8].

Recently, many studies have demonstrated that performance of composite material could be significantly improved by combination or copolymerization with a functional monomer containing nano-SiO_2. The composite material containing Nano-SiO_2, such as polyethylene terephthalate [9], styrene butadiene rubber [10], polyaniline [11], polyimide [12], and nylon 6 [13], may exhibit more satisfactory thermal stability, toughness, and strength owing to the effect of physical adsorption, hydrogen bond, Si–O bond, and C–Si bond [10, 12, 14–16]. However, there are no papers about the application of nano-SiO_2 in polymer for flooding to develop temperature tolerance, salt tolerance, and shear resistance of copolymer.

Keeping in mind these fundamental conditions, herein, a novel nano-SiO_2 functional monomer (NSFM; see Scheme 1) was introduced into AM/AA copolymer aiming to obtain satisfying temperature tolerance, salt tolerance, and shear resistance [17–20].

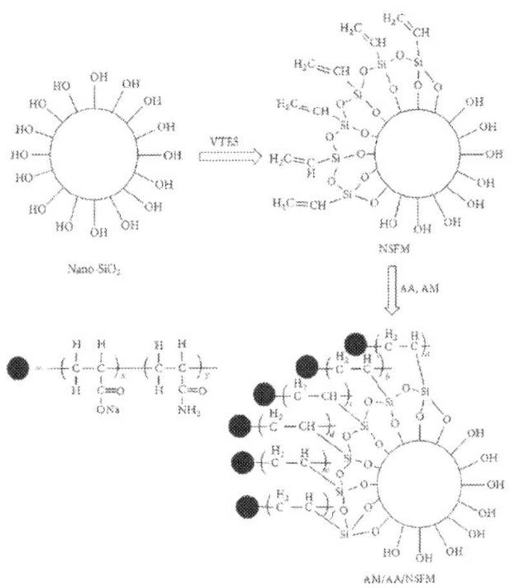

Scheme 1: The synthesis of AM/AA/NSFM.

EXPERIMENTAL

Chemicals and Reagents

Ethanol (C_2H_5OH, ≥99.7%), ammonia (NH_4OH, 28.0%), vinyltriethoxysilane (VTES, ≥98.0%), acrylic acid (AA, ≥99.5%), acrylamide (AM, ≥99.0%), sodium hydrogen sulfite ($NaHSO_3$, ≥58.5%), ammonium persulfate ($(NH_4)_2S_2O_8$, ≥98.0%), sodium hydroxide (NaOH, ≥96.0%), sodium chloride (NaCl, ≥99.5%), magnesium chloride hexahydrate ($MgCl_2{\cdot}6H_2O$, ≥98.0%), calcium chloride anhydrous ($CaCl_2$, ≥96.0%), potassium chloride (KCl, ≥99.5%), sodium sulfate (Na_2SO_4, ≥99.0%), and sodium bicarbonate ($NaHCO_3$, ≥99.5%) were purchased from Chengdu Kelong Chemical Reagent Factory (Sichuan, China). Nano-SiO_2 (10–20 nm) was obtained from Aladdin chemistry (Shanghai, China) Co., Ltd. All chemicals and reagents were used as received without any further purification. Water was deionized by passing through an ion-exchange column and doubly distilled.

Preparation of Nano-SiO_2 Functional Monomer

Firstly, 83.6 mL ethanol, 1.5 g nano-SiO_2, 13.6 mL distilled water, and 1.3 mL ammonia were added into a 250 mL round-bottom flask, and the mixture was dispersed with supersonic wave for 30 min. Then 2.0 mL VTES was added into the stirred solution in the round-bottom flask, and the reaction time was 18 h at 30°C. After reaction, the product was NSFM which was separated by centrifugation and washed with distilled water [21–23].

Synthesis of AM/AA/NSFM

0.05 g NSFM, 6.50 g AM, 3.45 g AA, and a certain amount of distilled water were added into a 100 mL three-necked flask, respectively, and the pH value of the mixture was regulated to 7.0 using sodium hydroxide solution; then the solution with 20% total monomer mass concentration was prepared. 0.05 g $NaHSO_3$-$(NH_4)_2S_2O_8$ initiator (mol ratio = 1 : 1) was taken along with distilled water in the three-necked flask assembled with a nitrogen (N_2) inlet. Then the copolymerization was carried out at 45°C under N_2 atmosphere for 6 h. Finally, the AM/AA/NSFM copolymer was obtained after ethanol washing, drying, and pulverizing. The synthesis of AM/AA/NSFM is shown in Scheme 1 [20]. The AM/AA copolymer was synthesized by using the same method.

Characterization

Infrared (IR) spectra of AM/AA and AM/AA/NSFM were measured with KBr pellets using a Perkin Elemer RX-1 spectrophotometer. ^{1}H NMR spectrum of AM/AA/NSFM was recorded on a Bruker AC-E 200 spectrometer by dissolving the copolymer in D_2O and operating at 400 MHz. The elementary analysis of AM/AA/NSFM was carried out with a Vario EL-III elemental analyzer. The microstructures of AM/AA and AM/AA/NSFM were observed by a scanning electron microscope (SEM). The weight-average molecular weight (M_w) of the copolymers was obtained by using a BI-200SM wide angle dynamic/static laser light scattering apparatus.

Intrinsic Viscosity

The intrinsic viscosity of copolymer was measured with an Ubbelohde viscometer at 25°C. The test temperature was controlled using a constant temperature bath. The flux time was reproducible to 0.05 s using a stopwatch. The copolymer solutions, at five different concentrations (0.1000, 0.0667, 0.0500, 0.0333, and 0.0250 wt%), were prepared with distilled water. The specific viscosity is calculated via the following equation [6, 24, 25]:

$$\eta_{sp} = \frac{t - t_0}{t_0}, \tag{1}$$

where η_{sp} is the specific viscosity of copolymer; t 0 is flux time of distilled water, s; and t is flux time of copolymer solution, s. Then the intrinsic viscosity is calculated with the Huggins equation [6, 26]:

$$\frac{\eta_{sp}}{C} = [\eta] + K'[\eta]^2 C, \tag{2}$$

where $[\eta]$ is intrinsic viscosity, mL/g; C is concentration of copolymer solution, wt%; and K' is the Huggins constant.

Rheological Property and Viscoelasticity

Rheological property and viscoelasticity measurements of the copolymers were conducted on a HAAKE RS 600 Rotational Rheometer (Germany). The shear rate was from 0.007 s^{-1} to 500 s^{-1}, and the temperature was 65°C with a heating rate of 1.5°C/min, while the test system was binocular tube and the rotor was DG41Ti in rheological measurements. The scanning range of frequency (f) was 0.01–10 Hz, and the stress was 0.1 Pa by using the same test system and rotor in viscoelasticity measurements.

Mobility Control Ability

The mobility control ability of the copolymer solutions is characterized by the resistance factor (RF) and the residual resistance factor (RRF) [27–29]. The RF is calculated with the following equation:

$$\mathrm{RF} = \frac{K_w/\mu_w}{K_p/\mu_p}, \tag{3}$$

where K_w is aqueous phase permeability, mD; K_p is polymer phase permeability, mD; μ_w is the viscosity of aqueous phase, mPa·s; and μ_p is the viscosity of polymer phase, mPa·s.

The RRF is calculated with the following equation:

$$\mathrm{RRF} = \frac{K_{\mathrm{wb}}}{K_{\mathrm{wa}}}, \tag{4}$$

where K_{wb} is aqueous phase permeability before polymer flooding, mD; K_{wa} is aqueous phase permeability after polymer flooding, mD.

Core Flooding Tests

Two Berea sandstone cores were used for core flooding experiments. The cores were dried at 65°C, and then their length, diameter, porosity, and gas permeability were measured by using a SCMS-B2 core multiparameter measurement system. A Hassler core holder was used with 3.5 MPa confining pressure and 1.5 MPa backpressure. The core, after being saturated with brine, was saturated with crude oil (52.5 mPas at 65°C) at 0.1-0.2 mL/min until irreducible water saturation S_{wi} was obtained. After 72 h of aging, the core was flooded by the brine at 0.2 mL/min until water cut was up to 95%, and then the copolymer solution (0.2 wt%) was injected at 0.2 mL/min until water cut reached 95% once more [6, 29]. All the core flooding procedures were conducted at 65°C. Chemical composition and total dissolved solids (TDS) of the brine are listed in Table 1. The maximum work pressure of the ISCO 260D syringe pump is 50 MPa, and its minimum and maximum displacement velocity is 0.001 and 50.000 mL/min, respectively. The EOR is calculated with the following equation:

$$\mathrm{EOR} = E - E_w, \tag{5}$$

where EOR is enhanced oil recovery, %; E is the oil recovery of the whole displacement process, %; and E_w is the oil recovery of water flooding, %.

Table 1: Composition and TDS of brine

Composition	NaCl	KCl	$CaCl_2$	$MgCl_2 6H_2O$	Na_2SO_4	$NaHCO_3$	TDS
Content (wt%)	0.8495	0.0149	0.0764	0.2154	0.0125	0.0428	1.0951

Flow chart of the core flooding experiments is shown in Figure 1.

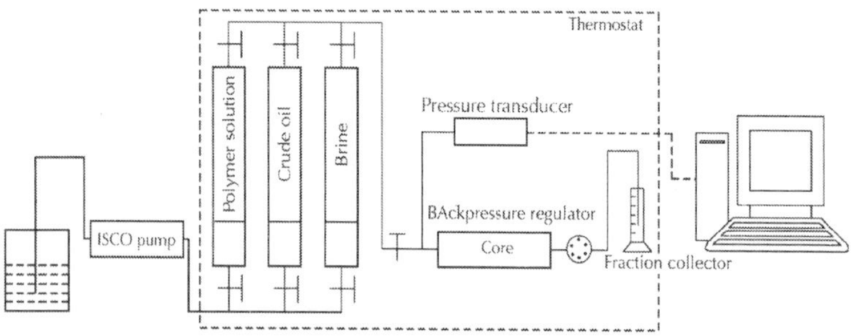

Figure 1: Flow chart of the core flooding experiments.

RESULTS AND DISCUSSION

IR Spectra Analysis

The structures of AM/AA and AM/AA/NSFM were confirmed by IR spectra as illustrated in Figure 2. The AM/AA/NSFM which was prepared using acrylic acid, acrylamide, and NSFM by free radical polymerization was confirmed by strong absorptions at 3419 cm^{-1} (–NH stretching vibration and –OH stretching vibration), 2942 cm^{-1} (–CH_2 stretching vibration), 1675 cm^{-1} (C=O stretching vibration), 1402 cm^{-1} (C–N stretching vibration), 1100 cm^{-1} (Si–O–Si asymmetric stretching vibration), and 780 cm^{-1} (Si–O–Si symmetric stretching vibration) in the spectrum of AM/AA/NSFM [18, 20]. The peak at 3419 cm^{-1} was broad in the IR spectrum of AM/AA/NSFM partly due to the hydroxyl on nano-SiO_2 surface [20]. As expected, the IR spectra confirmed the presence of different monomers in AM/AA/NSFM.

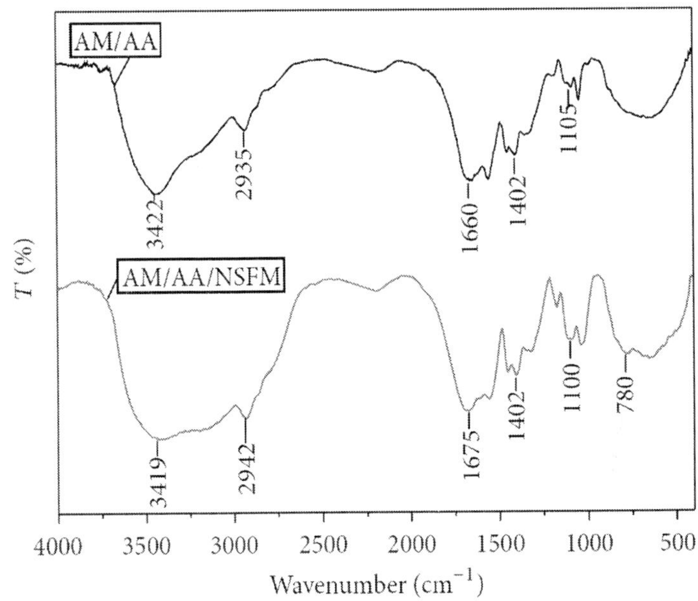

Figure 2: IR spectra of AM/AA and AM/AA/NSFM.

^{1}H NMR Analysis

The ^{1}H NMR spectrum of AM/AA/NSFM is shown in Figure 3. The chemical shift value at 1.29 ppm is due to the protons of [–CH_2–CH (Si(O–)$_3$)–]. The chemical shift value at 1.61 ppm is assigned to the protons of [–CH_2–CH ($CONH_2$)–] and [–CH_2–CH (COONa)–]. The protons of [–CH_2–CH ($CONH_2$)–] and [–CH_2–CH(COONa)–] appear at 2.17 ppm. The characteristic peak due to the protons of [–CH_2–CH (Si(O–)$_3$)–] is observed at 2.38 ppm.

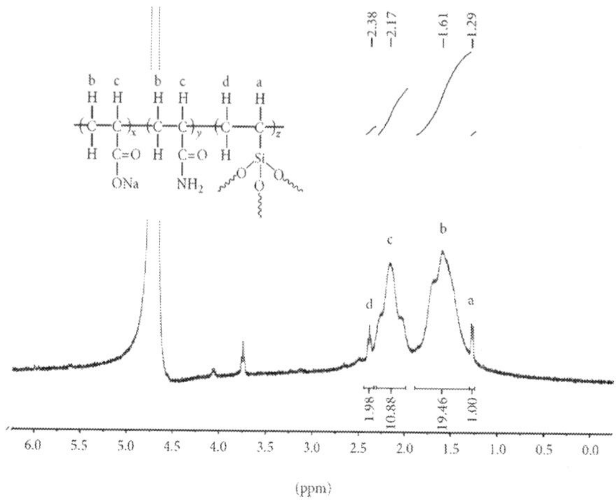

Figure 3: ^{1}H NMR spectrum of AM/AA/NSFM in D_2O.

Elementary Analysis of AM/AA/NSFM

The elementary analysis of the AM/AA/NSFM copolymer was carried out using a Vario EL-III elemental analyzer. The content of different element in the copolymer can be obtained by detecting the gases, which are the decomposition products of the copolymer at high temperature. Theoretical value: 0.21% (Si %), 45.4% (C %), and 5.4% (H %); found value: 0.17% (Si %), 40.1% (C %), and 4.8% (H %).

Microscopic Structure

The microscopic structures of AM/AA and AM/AA/NSFM were observed through SEM at room temperature. The copolymers solution samples (0.05 wt%) were prepared with distilled water and cooled with liquid nitrogen, and then these samples were evacuated in order to keep original appearance of the copolymers as far as possible. As shown in Figures 4 and 5, the molecular chains of copolymer were obviously changed when NSFM was introduced into the AM/AA copolymer. Compared with the images of AM/AA, the molecular coils of AM/AA/NSFM were composed of many micro-nano structure units, and the force between these units could be heightened due to Si–O and C–Si bonds. In addition, this structure may increase retention of AM/AA/NSFM on the rock face which is favorable to mobility control and EOR.

(a)

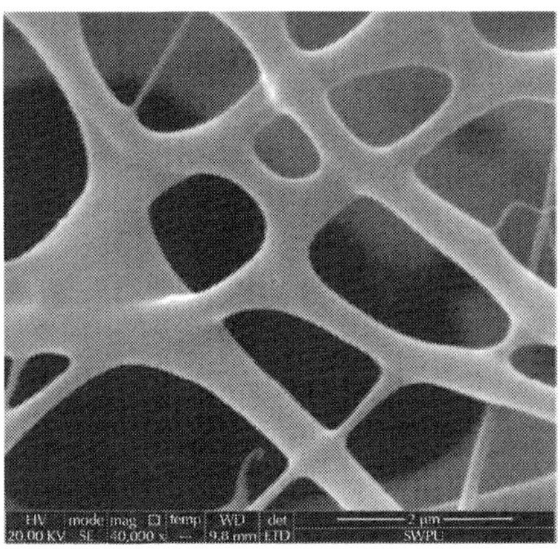

(b)

Figure 4: SEM images of AM/AA.

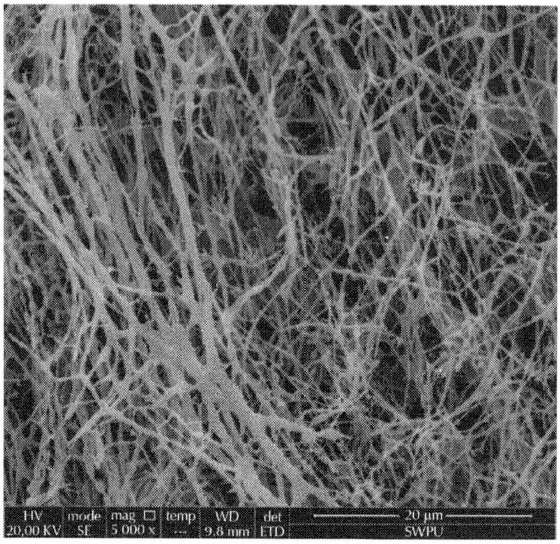

(a)

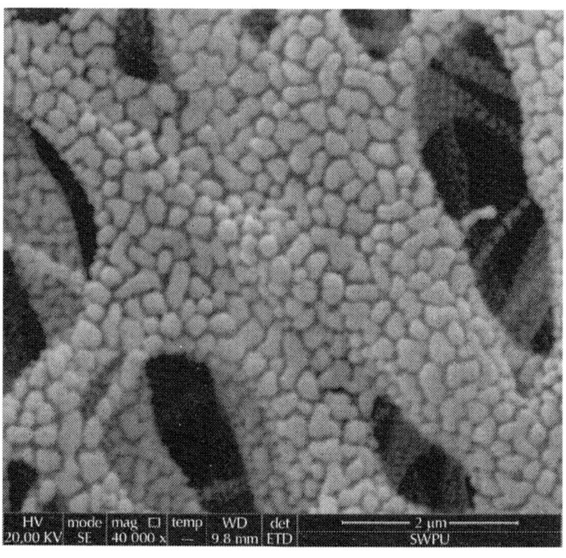

(b)

Figure 5: SEM images of AM/AA/NSFM.

Weight-Average Molecular Weight

Five different concentrations (0.001, 0.002, 0.004, 0.006, and 0.008 wt%) copolymer solutions were prepared with distilled water and filtered using a 0.5 μm Millipore Millex-LCR filter before static laser light scattering (SLLS) experiments. The M_w of AM/AA and AM/AA/NSFM can be calculated with the following equation [30]:

$$\frac{KC}{R_{vv}(q)} \cong \frac{1}{M_w}\left(1 + \frac{1}{3}\langle R_g \rangle^2 q^2\right), \tag{6}$$

where K is a constant; C is the concentration of copolymer solution, g/mL; $R_{vv}(q)$ is the Rayleigh ratio; $\langle R_g \rangle$ is the average radius of gyration, nm; and $q = (4\pi n/\lambda_o)\sin(\theta/2)$ with θ, λ_o, and n being the scattering angle, the wavelength of light in vacuo, and the solvent refractive index, respectively.

The M_w of AM/AA and AM/AA/NSFM is $(1.33 \pm 0.30) \times 10^7$ g/mol and $(1.32 \pm 0.45) \times 10^7$ g/mol, respectively (for details, see Supporting Material available online at http:// dx.doi.org/10.1155/2013/437309).

Intrinsic Viscosity

The η_{sp}/C versus C relationship is shown in Figure 6. The fitted line of η_{sp}/C versus C wasextrapolated to zero concentration.According to theHuggins equation, the y-intercept is the intrinsic viscosity of the copolymers. The results revealed that the intrinsic viscosity of AM/AA and AM/AA/NSFM was 733.9 and 789.5 mL/g, respectively.

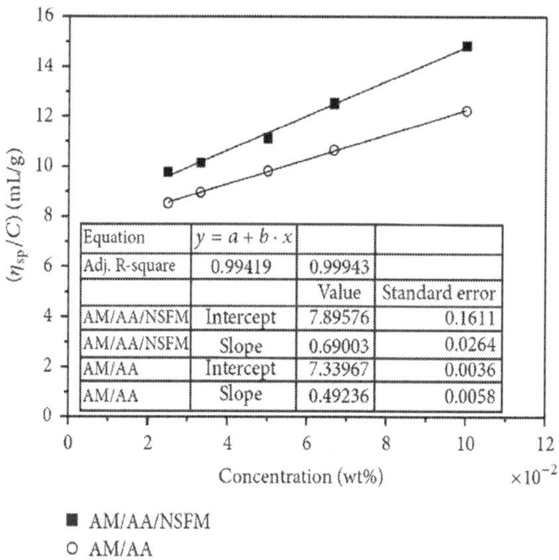

Figure 6: The η_{sp}/C versus C relationships of AM/AA and AM/AA/NSFM.

Shear Resistance

The viscosity versus shear rate curves of AM/AA and AM/AA/NSFM (0.2 wt%) are shown in Figure 7(a). It was clearly found that AM/AA and AM/AA/NSFM revealed non-Newtonian shear-thinning behavior. Hence, with the increase of the shear rate (from 0.007 to 500 s^{-1}), the viscosity of copolymer solutions dropped obviously. The results indicated that AM/AA/NSFM had better viscosifying property than AM/AA, and the viscosity of AM/AA/NSFM was higher than that of AM/AA at 500 s^{-1} shear rate (18.6 mPas versus 8.7 mPas). Furthermore, AM/AA and AM/AA/NSFM were investigated by changing the shear rate from 124 s^{-1} to 500 $s^{-1}$and from 500 s^{-1} to 124 s^{-1} around (Figures 7(b) and 7(c)). Compared with AM/AA, AM/AA/NSFM had higher retention rate of viscosity (85% versus 68%) when one cycle was completed? This phenomenon may support the Si–O and C–Si bonds in AM/AA/NSFM which can improve the shear tolerance of the copolymer. The structures of AM/AA/NSFM may be restored after being sheared.

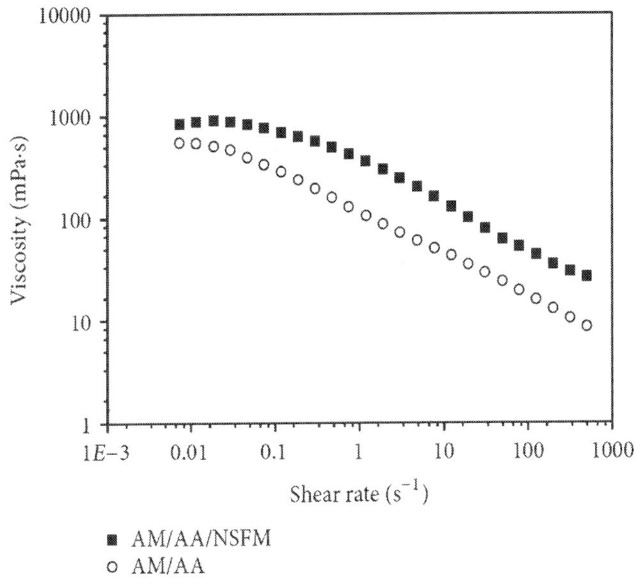
Viscosity (mPa·s)
10000
1000
100
10
1
1E−3
0.01
0.1
1
10
100
1000
Shear rate (s−1)
AM/AA/NSFM
AM/AA

(a)

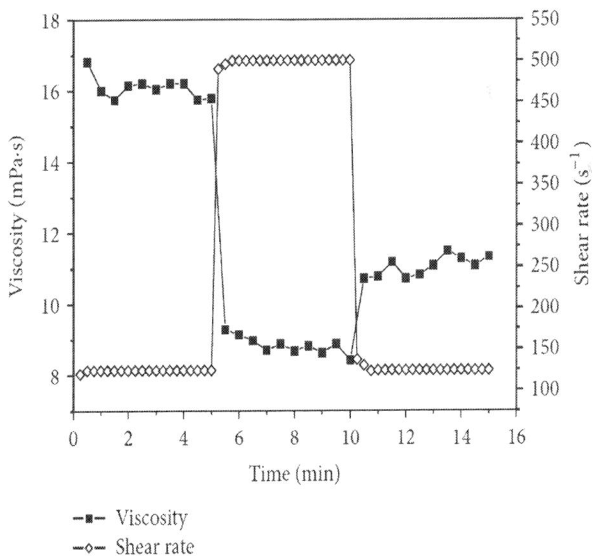
Viscosity (mPa·s)
Shear rate (s−1)
Time (min)
Viscosity
Shear rate

(b)

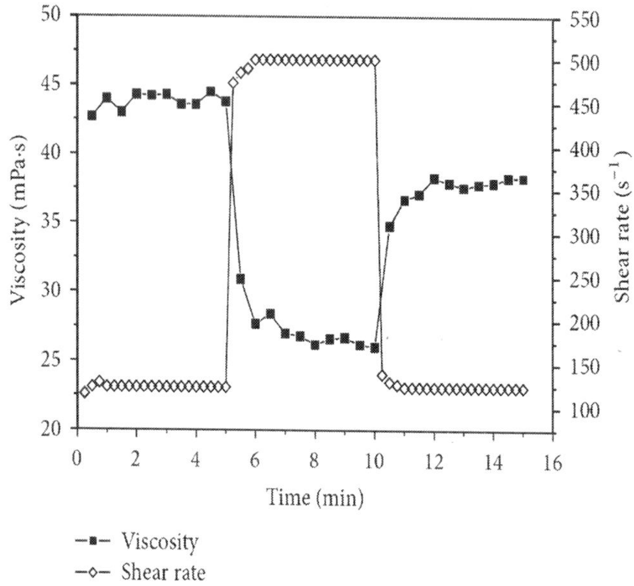

(c)

Figure 7: (a) Effect of shear rate on viscosity; (b) shear resistance of AM/AA; (c) shear resistance of AM/AA/NSFM. The copolymers solutions (0.2 wt%) were prepared with distilled water.

Viscoelasticity Measurements

The viscoelasticity curves of AM/AA and AM/AA/NSFM solutions (0.2 wt%) are shown in Figure 8. When the frequency was lower than 1 Hz, the viscous modulus (G″) of AM/AA/NSFM was higher than the elastic modulus (G′); when the frequency was higher than 1 Hz, the situation was just the opposite. However, the G″ of AM/AA was higher than G′ in the entire frequency scan range. Compared with AM/AA, AM/AA/NSFM exhibited higher G′ and G″ under the same conditions. This phenomenon may support the micro-nano structure units in AM/AA/NSFM can enhance the acting force of polymer molecular coils.

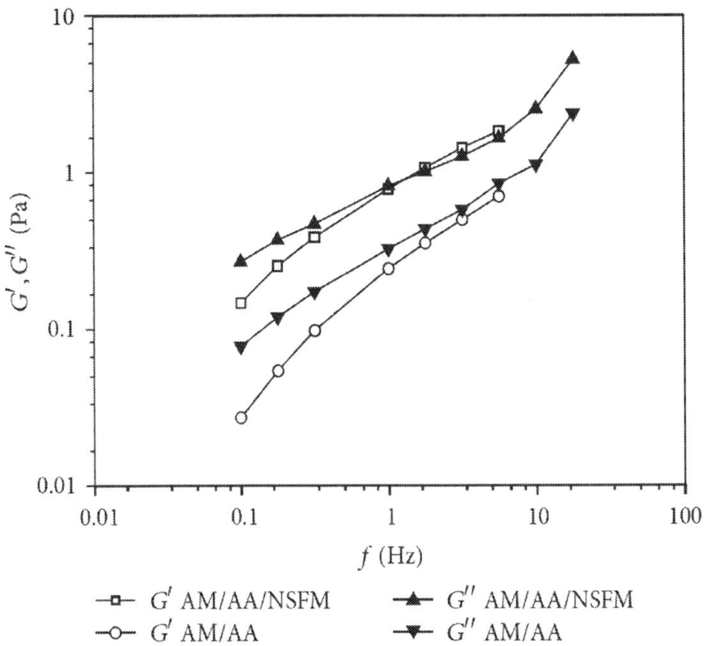

Figure 8: Viscoelasticity of AM/AA and AM/AA/NSFM at 65°C. The copolymers solutions (0.2 wt%) were prepared with distilled water.

Temperature Tolerance

AM/AA and AM/AA/NSFM solutions were prepared with distilled water. And the viscosity of copolymer solutions was measured by the Brookfiled DV-3 viscometer at different temperatures. The viscosity versus temperature curves of AM/AA and AM/AA/NSFM solutions are shown in Figure 9. The test results showed that the AM/AA/NSFM solution had higher viscosity at the same temperature. Additionally, the viscosity of AM/AA/NSFM solution decreased less than that of AM/AA when temperature was above 80°C. This may support the stable Si–O and C–Si bonds which can obviously improve temperature tolerance of AM/AA/NSFM.

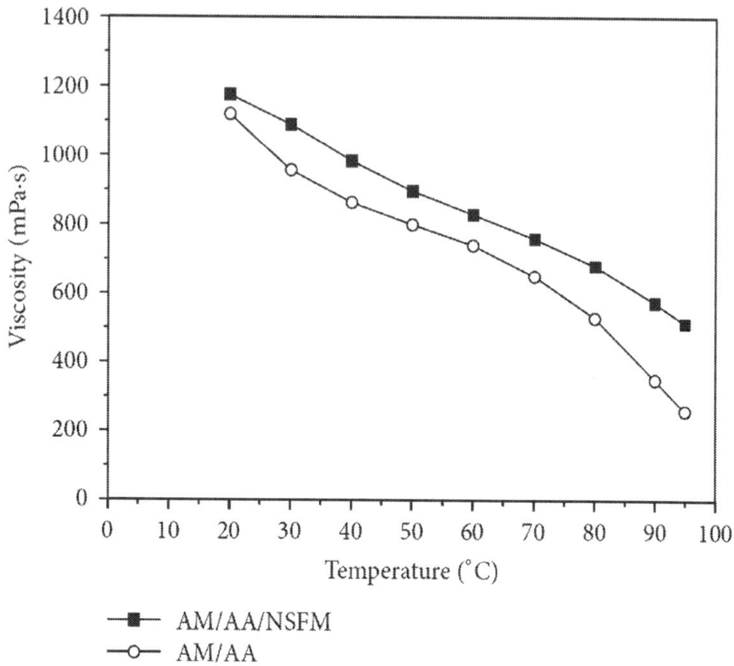

Figure 9: Viscosity versus temperature for AM/AA and AM/AA/NSFM solution. The viscosity of copolymer solution (0.5 wt%) was measured by Brookfiled DV-3 viscometer at 7.34 s^{-1} using number 62 rotor (rotation speed: 18.8 r/min).

Salt Tolerance

As shown in Figures 10(a), 10(b), and 10(c), with the increase of salt concentration (NaCl, $CaCl_2$, and $MgCl_2 6H_2O$), the viscosity of copolymers decreases rapidly and then kept at a low value. It was found that AM/AA and AM/AA/NSFM had less satisfactory salt tolerance to Na^+ or Ca^{2+} than to Mg^{2+} under the same conditions. Compared with AM/AA, AM/AA/NSFM exhibited no obvious advantage in salt tolerance due to the shrinking of copolymer chain with the increase of salt concentration.

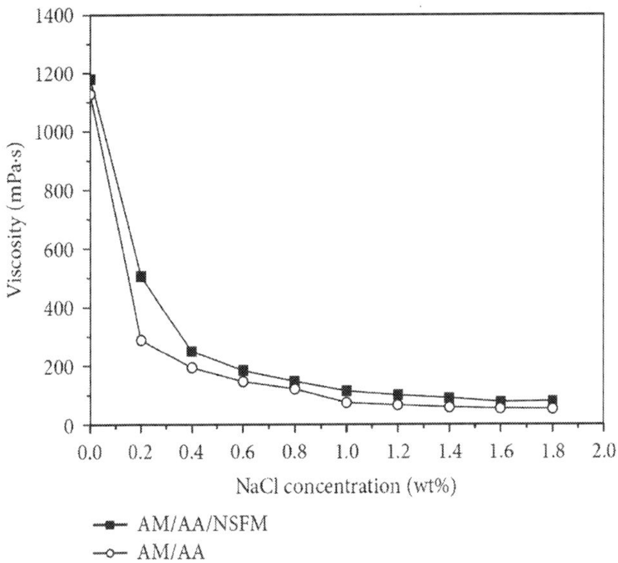
1400
1200
1000
800
600
400
200
0
Viscosity (mPa·s)
0.0 0.2 0.4 0.6 0.8 1.0 1.2 1.4 1.6 1.8 2.0
NaCl concentration (wt%)
AM/AA/NSFM
AM/AA

(a)

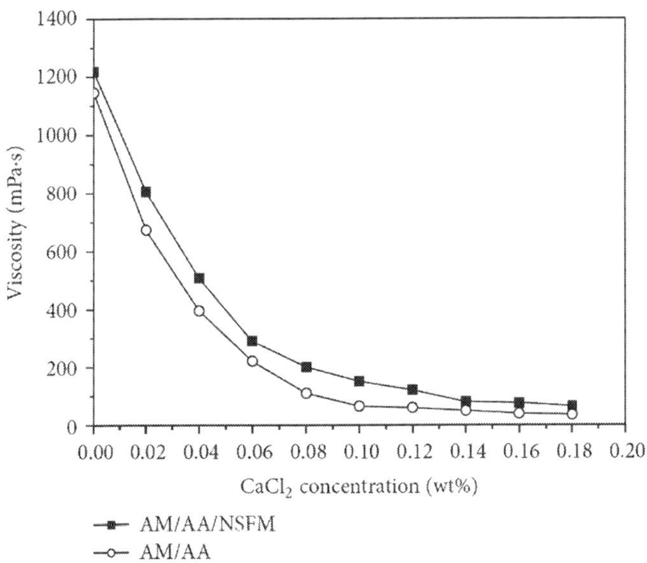
1400
1200
1000
800
600
400
200
0
Viscosity (mPa·s)
0.00 0.02 0.04 0.06 0.08 0.10 0.12 0.14 0.16 0.18 0.20
CaCl2 concentration (wt%)
AM/AA/NSFM
AM/AA

(b)

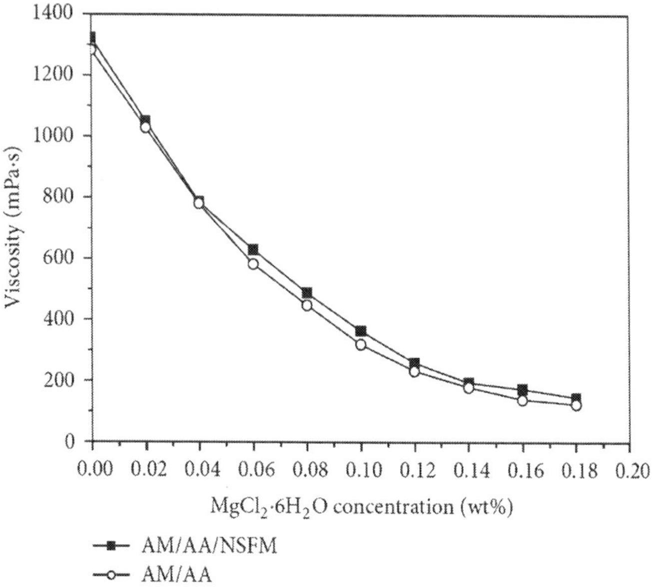

(c)

Figure 10: Salt tolerance ((a) NaCl, (b) $CaCl_2$, and (c) $MgCl_2 6H_2O$) of AM/AA and AM/AA/NSFM solutions (0.5 wt%) at 20°C. The viscosity of copolymer solution was measured by Brookfiled DV-3 viscometer at 7.34 s^{-1} using number 62 rotor (rotation speed: 18.8 r/min) or number 61 rotor (rotation speed: 18.5 r/min).

Mobility Control Ability

The core barrel was packed with quartz sand which was washed by hydrochloric acid and distilled water for several times. The injection rate of brine (sodium chloride concentration was 0.5 wt%) and polymer solution prepared with the brine was 2.0 mL/min with the ISCO 260D syringe pump. Experiments were carried out at 65°C in an incubator with precision of 0.1°C. The injection pressure was collected by a pressure sensor with precision of 0.0001 MPa. The flow characteristic curves of AM/AA and AM/AA/NSFM in porous media are shown in Figure 11.

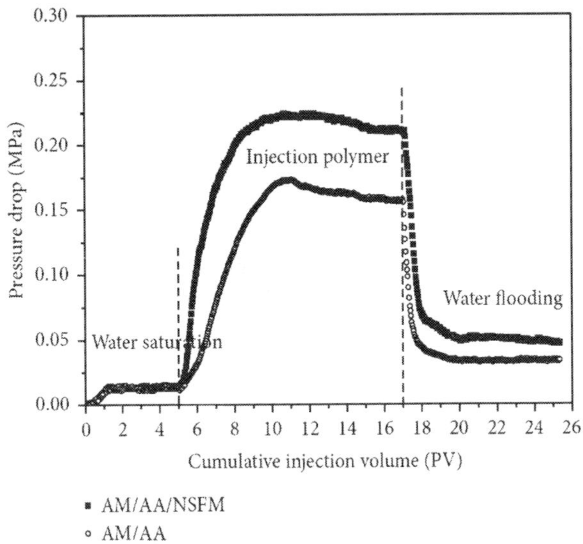

Figure 11: Flow characteristic curves of AM/AA and AM/AA/NSFM solution (0.2 wt%). The length and internal diameter of the core barrel were 25.0 cm and 2.5 cm, respectively.

As shown in Figure 11, the AM/AA/NSFM solution could establish much higher RF and RRF than that of the AM/AA solution under the same conditions (RF: 16.52 versus 12.17, RRF: 3.63 versus 2.59). This is to say that the AM/AA/NSFM solution has stronger mobility control ability which is favorable to enhance oil recovery due to the higher viscosity retention rate and microstructure. In addition, it was found that AM/AA/NSFM revealed higher retention than AM/AA (83 mg versus 55 mg) by material balance calculations (for details, see Supporting Materials). This may support that the huge surface area of micro-nano structure units of AM/AA/NSFM can enhance the adsorption which may play an important role in improving mobility control.

Enhanced Oil Recovery

As shown in Table 2, the EOR of AM/AA/NSFM solution (0.2 wt%) was 20.10% compared with water flooding at 65°C. However, the

EOR of AM/AA solution (0.2 wt%) was 14.22% under the same conditions. The EOR results showed that AM/AA/NSFM revealed more superior ability of oil displacement. As shown in Figure 12, compared with AM/AA, AM/AA/NSFM exhibited stronger ability of reducing water cut and establishing flow resistance in polymer flooding process. This phenomenon may support that the sweep efficiency is obviously improved by AM/AA/NSFM due to the excellent mobility control capability in porous media.

Table 2: The relevant core properties and the results of core flooding experiments

Copolymer	**Cores**	**Length (cm)**	**Diameter (cm)**	**Porosity (%)**	**Permeability (mD)**	S_{wi} (%)	**E (%)**	E_w (%)	**EOR (%)**
AM/AA	1#	8.84	3.77	23.11	937.03	19.18	45.73	31.51	14.22
AM/AA-/ NSFM	2#	8.92	3.78	23.05	926.28	19.37	52.85	31.75	20.10

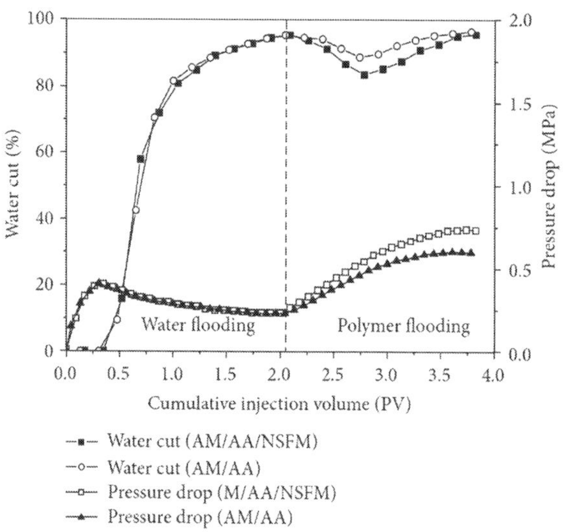

Figure 12: Core flooding experiments results of AM/AA/NSFM and AM/AA (0.2 wt%) at 65°C.

CONCLUSIONS

A novel copolymer containing nano-SiO_2 was synthesized by free radical polymerization using AM, AA, and NSFM as raw materials. The AM/AA/NSFM copolymer was characterized by IR spectrum, ^1H NMR spectrum, elemental analysis, and scanning electron microscope. The solution properties, such as rheological property, viscoelasticity, temperature tolerance, salt tolerance, mobility control ability, and oil displacement efficiency of the copolymer, were investigated under different conditions. The results indicated that the copolymer containing nano-SiO_2 possessed moderate or good shear resistance, temperature tolerance, and mobility control ability as EOR chemical.

ACKNOWLEDGMENTS

This work was supported by the Open Fund (PLN1212) of State Key Laboratory of Oil and Gas Reservoir Geology and Exploitation (Southwest Petroleum University) and the Specialized Research Fund for the Doctoral Program of Higher Education (20125121120011).

REFERENCES

1. N. Mungan, F. W. Smith, J. L. Thompson, O. Sinclair, and C. Gas, "Some aspects of polymer floods,"Journal of Petroleum Technology, vol. 18, no. 9, pp. 1143–1150, 1966.
2. B. S. Shiran and A. Skauge, "Enhanced oil recovery (EOR) by combined low salinity water/polymer flooding," Energy Fuels, vol. 27, no. 3, pp. 1223–1235, 2013.
3. C. Zhong, R. Huang, X. Zhang, and H. Dai, "Synthesis, characterization, and solution properties or an acrylamide-based terpolymer with butyl styrene," Journal of Applied Polymer Science, vol. 103, no. 6, pp. 4027–4038, 2007.

4. T. Rho, J. Park, C. Kim, H. Yoon, and H. Suh, "Degradation of polyacrylamide in dilute solution,"Polymer Degradation and Stability, vol. 51, no. 3, pp. 287–293, 1996.
5. D. A. Z. Wever, F. Picchioni, and A. A. Broekhuis, "Polymers for enhanced oil recovery: a paradigm for structure-property relationship in aqueous solution," Progress in Polymer Science, vol. 36, no. 11, pp. 1558–1628, 2011.
6. Z. B. Ye, G. J. Gou, S. H. Gou, W. C. Jiang, and T. Y. Liu, "Synthesis and characterization of a water-soluble sulfonates copolymer of acrylamide and N-Allylbenzamide as enhanced oil recovery chemical,"Journal of Applied Polymer Science, vol. 128, no. 3, pp. 2003–2011, 2013.
7. S. H. Chang and I. J. Chung, "Effect of shear flow on polymer desorption and latex dispersion stability in the presence of adsorbed polymer," Macromolecules, vol. 24, no. 2, pp. 567–571, 1991. ·
8. L. Xue, U. S. Agarwal, and P. J. Lemstra, "Shear degradation resistance of star polymers during elongational flow," Macromolecules, vol. 38, no. 21, pp. 8825–8832, 2005. · ·
9. J. Zheng, P. Cui, X. Tian, and K. Zheng, "Pyrolysis studies of polyethylene terephthalate/silica nanocomposites," Journal of Applied Polymer Science, vol. 104, no. 1, pp. 9–14, 2007. ·
10. L. I. Rueda, L. G. Hernandez, and C. C. Anton, "Effect of the textural characteristics of the new silicas on the dynamic properties of Styrene-Butadiene Rubber (SBR) vulcanizates," Polymer Composites, vol. 9, no. 3, pp. 204–208, 1988. ·
11. H. Xia and Q. Wang, "Preparation of conductive polyaniline/nanosilica particle composites through ultrasonic irradiation," Journal of Applied Polymer Science, vol. 87, no. 11, pp. 1811–1817, 2003.
12. X. Wang, X. Zhao, M. Wang, and Z. Shen, "The effects of atomic oxygen on polyimide resin matrix composite containing nano-silicon dioxide," Nuclear Instruments and Methods in Physics Research B, vol. 243, no. 2, pp. 320–324, 2006.

13. Y. Li, J. Yu, and Z. Guo, "The influence of interphase on nylon-6/nano-SiO_2 composite materials obtained from in situ polymerization," Polymer International, vol. 52, no. 6, pp. 981–986, 2003.
14. V. A. Bershtein, L. M. Egorova, P. N. Yakushev, P. Pissis, P. Sysel, and L. Brozova, "Molecular dynamics in nanostructured polyimide-silica hybrid materials and their thermal stability," Journal of Polymer Science B, vol. 40, no. 10, pp. 1056–1069, 2002.
15. N. D. Alberola, K. Benzarti, C. Bas, and Y. Bomal, "Interface effects in elastomers reinforced by modified precipitated silica," Polymer Composites, vol. 22, no. 2, pp. 312–325, 2001.
16. A. Voronov, A. Kohut, A. Synytska, and W. Peukert, "Mechanochemical modification of silica with poly(1-vinyl-2-pyrrolidone) by grinding in a stirred media mill," Journal of Applied Polymer Science, vol. 104, no. 6, pp. 3708–3714, 2007.
17. H. Zou, S. Wu, and J. Shen, "Polymer/silica nanocomposites: preparation, characterization, propertles, and applications," Chemical Reviews, vol. 108, no. 9, pp. 3893–3957, 2008.
18. G. Hsiue, W. Kuo, Y. Huang, and R. Jeng, "Microstructural and morphological characteristics of PS-SiO_2 nanocomposites," Polymer, vol. 41, no. 8, pp. 2813–2825, 2000.
19. W. Wang and B. Gu, "Self-assembly of two- and three-dimensional particle arrays by manipulating the hydrophobicity of silica nanospheres," Journal of Physical Chemistry B, vol. 109, no. 47, pp. 22175–22180, 2005.
20. B. J. Kim and K. S. Kang, "Fabrication of a crack-free large area photonic crystal with colloidal silica spheres modified with vinyltriethoxysilane," Crystal Growth & Design, vol. 12, no. 8, pp. 4039–4042, 2012.
21. A. V. Biradar, A. A. Biradar, and T. Asefa, "Silica-dendrimer core-shell microspheres with encapsulated ultrasmall palladium nanoparticles: efficient and easily recyclable

heterogeneous nanocatalysts,"Langmuir, vol. 27, no. 23, pp. 14408–14418, 2011.

22. Z. Meng, C. Xue, Q. Zhang, X. Yu, K. Xi, and X. Jia, "Preparation of highly monodisperse hybrid silica nanospheres using a one-step emulsion reaction in aqueous solution," Langmuir, vol. 25, no. 14, pp. 7879–7883, 2009.
23. Z. Wu, H. Han, W. Han, B. Kim, K. H. Ahn, and K. Lee, "Controlling the hydrophobicity of submicrometer silica spheres via surface modification for nanocomposite applications," Langmuir, vol. 23, no. 14, pp. 7799–7803, 2007.
24. A. Mehrdad and R. Akbarzadeh, "Effect of temperature and solvent composition on the intrinsic viscosity of poly(vinyl pyrrolidone) in water-ethanol solutions," Journal of Chemical and Engineering Data, vol. 55, no. 9, pp. 3720–3724, 2010.
25. J. Lee and A. Tripathi, "Intrinsic viscosity of polymers and biopolymers measured by microchip,"Analytical Chemistry, vol. 77, no. 22, pp. 7137–7147, 2005.
26. L. Alagha, S. Wang, Z. Xu, and J. Masliyah, "Adsorption kinetics of a novel organic-inorganic hybrid polymer on silica and alumina studied by quartz crystal microbalance," Journal of Physical Chemistry C, vol. 115, no. 31, pp. 15390–15402, 2011.
27. R. Ponnapati, O. Karazincir, E. Dao, R. Ng, K. K. Mohanty, and R. Krishnamoorti, "Polymer-functionalized nanoparticles for improving waterflood sweep efficiency: characterization and transport properties," Industrial and Engineering Chemistry Research, vol. 50, no. 23, pp. 13030–13036, 2011.
28. C. Zhong, L. Ye, H. Dai, and R. Huang, "Flourescent probe and ESEM morphologies of a acrylamide-based terpolymer in aqueous solution," Journal of Applied Polymer Science, vol. 103, no. 1, pp. 277–286, 2007.
29. L. Shi, Z. Ye, Z. Zhang, C. Zhou, S. Zhu, and Z. Guo, "Necessity and feasibility of improving the residual resistance factor of

polymer flooding in heavy oil reservoirs," Petroleum Science, vol. 7, no. 2, pp. 251–256, 2010.

30. X. Wang, X. Qiu, and C. Wu, "Comparison of the coil-to-globule and the globule-to-coil transitions of a single poly(N-isopropylacrylamide) homopolymer chain in water," Macromolecules, vol. 31, no. 9, pp. 2972–2976, 1998.

Chapter 3

A Review on Heavy Metals (As, Pb, and Hg) Uptake by Plants through Phytoremediation

Bieby Voijant Tangahu[1], Siti Rozaimah Sheikh Abdullah[2], Hassan Basri[1], Mushrifah Idris[3], Nurina Anuar[2], and Muhammad Mukhlisin[1]

[1]Department of Civil and Structural Engineering, Faculty of Engineering and Built Environment, Universiti Kebangsaan Malaysia, 43600 Bangin, Malaysia

[2]Department of Chemical Engineering, Faculty of Engineering and Built Environment, Universiti Kebangsaan Malaysia, 43600 Bangin, Malaysia

[3]Tasik Chini Reasearch Centre, Faculty of Science and Technology, Universiti Kebangsaan Malaysia, 43600 Bangin, Malaysia

ABSTRACT

Heavy metals are among the most important sorts of contaminant in the environment. Several methods already used to clean up the environment from these kinds of contaminants, but most of them are costly and difficult to get optimum results. Currently, phytoremediation is an effective and affordable technological solution used to extract or remove inactive metals and metal pollutants from contaminated soil and water. This technology is environmental friendly and potentially cost effective. This paper aims to compile some information about heavy metals of arsenic, lead, and mercury (As, Pb, and Hg) sources, effects and their treatment. It also reviews deeply about phytoremediation technology, including the heavy metal uptake mechanisms and several research studies associated about the topics. Additionally, it describes several sources and the effects of As, Pb, and Hg on the environment, the advantages of this kind of technology for reducing them, and also heavy metal uptake mechanisms in phytoremediation technology as well as the factors affecting the uptake mechanisms. Some recommended plants which are commonly used in phytoremediation and their capability to reduce the contaminant are also reported.

INTRODUCTION

Heavy metals are among the contaminants in the environment. Beside the natural activities, almost all human activities also have potential contribution to produce heavy metals as side effects. Migration of these contaminants into noncontaminated areas as dust or leachates through the soil and spreading of heavy metals containing sewage sludge are a few examples of events contributing towards contamination of the ecosystems [1].

Several methods are already being used to clean up the environment from these kinds of contaminants, but most of them are costly and far away from their optimum performance. The chemical technologies generate large volumetric sludge and increase the costs [2]; chemical and thermal methods are both technically difficult and expensive that all of these methods can also degrade the valuable component of soils [3]. Conventionally, remediation of heavy-metal-contaminated soils involves either onsite management or excavation and subsequent disposal to a landfill site. This method of disposal solely shifts the contamination problem elsewhere along with the hazards associated with transportation of contaminated soil and migration of contaminants from landfill into an adjacent environment. Soil washing for removing contaminated soil is an alternative way to excavation and disposal to landfill. This method is very costy and produces a residue rich in heavy metals, which will require further treatment. Moreover, these physio-chemical technologies used for soil remediation render the land usage as a medium for plant growth, as they remove all biological activities [1].

Recent concerns regarding the environmental contamination have initiated the development of appropriate technologies to assess the presence and mobility of metals in soil [4], water, and wastewater. Presently, phytoremediation has become an effective and affordable technological solution used to extract or remove inactive metals and metal pollutants from contaminated soil. Phytoremediation is the use of plants to clean up a contamination from soils, sediments, and water. This technology is environmental friendly and potentially costeffective. Plants with exceptional metal-accumulating capacity are known as hyperaccumulator plants [5]. Phytoremediation takes the advantage of the unique and selective uptake capabilities of plant root systems, together with the translocation, bioaccumulation, and contaminant degradation abilities of the entire plant body [3].

Many species of plants have been successful in absorbing contaminants such as lead, cadmium, chromium, arsenic, and various radionuclides from soils. One of phytoremediation categories, phytoextraction, can be used to remove heavy metals from soil using its ability to uptake metals which are essential for plant growth (Fe, Mn, Zn, Cu, Mg, Mo, and Ni). Some metals with unknown biological function (Cd, Cr, Pb, Co, Ag, Se, Hg) can also be accumulated [5].

The objectives of this paper are to discuss the potential of phytoremediation technique on treating heavy metal-contaminated side, to provide a brief view about heavy metals uptake mechanisms by plant, to give some description about the performance of several types of plants to uptake heavy metals and to describe about the fate of heavy metals in plant tissue, especially on arsenic (As), lead (Pb), and mercury (Hg). This study is related to a research project that aims to identify potential plants in tropical country such as Malaysia which can uptake heavy metal contaminants from petrochemical wastewater.

HEAVY METALS: SOURCES AND EFFECT IN THEENVIRONMENT

Heavy metals are conventionally defined as elements with metallic properties and an atomic number >20. The most common heavy metal contaminants are Cd, Cr, Cu, Hg, Pb, and Zn. Metals are natural components in soil [6]. Some of these metals are micronutrients necessary for plant growth, such as Zn, Cu, Mn, Ni, and Co, while others have unknown biological function, such as Cd, Pb, and Hg [1].

Metal pollution has harmful effect on biological systems and does not undergo biodegradation. Toxic heavy metals such as Pb, Co, Cd can be differentiated from other pollutants, since they cannot be biodegraded but can be accumulated in living organisms, thus causing various diseases and disorders even in relatively lower concentrations [7]. Heavy metals, with soil residence times of thousands of years, pose numerous health dangers to higher organisms. They are also known to have effect on plant growth, ground cover and have a negative impact on soil microflora [8]. It is well known that heavy metals cannot be chemically degraded and need to be physically removed or be transformed into nontoxic compounds [1].

Arsenic (As)

Arsenic (atomic number 33) is a silver-grey brittle crystalline solid with atomic weight of 74.9, specific gravity 5.73, melting point 817°C (at 28 atm), boiling point 613°C, and vapor pressure 1 mm Hg at 372°C [9]. Arsenic is a semimetallic element with the chemical symbol "As". Arsenic is odorless and tasteless. Arsenic can combine with other elements to form inorganic and organic arsenicals [10]. In the environment, arsenic is combined with oxygen, chlorine, and sulfur to form inorganic arsenic compounds. Inorganic arsenic compounds are mainly used to preserve wood. Organic arsenic compounds are used as pesticides, primarily on cotton plants [11].

Arsenic exists in the −3, 0, +3, and +5 valence oxidation states [9], and in a variety of chemical forms in natural waters and sediments [12]. Environmental forms include arsenious acids (H_3AsO_3, H_3AsO_3, $H_3AsO_3^{2-}$), arsenic acids (H_3AsO_4, $H_3AsO_4^{-}$, $H_3AsO_4^{2-}$), arsenites, arsenates, methylarsenic acid, dimethylarsinic acid, and arsine. Two most common forms in natural waters arsenite (AsO_3^{3-}) and inorganic arsenate (AsO_4^{3-}), referred as As^{3+} and As^{5+} [9]. From both the biological and the toxicological points of view, arsenic compounds can be classified into three major groups. These groups are inorganic arsenic compounds, organic arsenic compounds, and arsine gas [13].

It is a hard acid and preferentially complexes with oxides and nitrogen. Trivalent arsenites predominate in moderately reducing anaerobic environments such as groundwater [9]. The most common trivalent inorganic arsenic compounds are arsenic trioxide, sodium arsenite, and arsenic trichloride [13]. Trivalent (+3) arsenates include As $(OH)_3$, $As(OH)_4^{-}$, AsO_2OH^{2-}, and AsO_3^{3-} [9]. Arsenite (As $(OH)_3$, As^{3+}) is predominant in reduced redox potential conditions [12].

Arsenic is one of the contaminants found in the environment which is notoriously toxic to man and other living organisms [14].

It is a highly toxic element that exists in various species, and the toxicity of arsenic depends on its species. The pH, redox conditions, surrounding mineral composition, and microbial activities affect the form (inorganic or organic) and the oxidation state of arsenic. It is generally accepted that the inorganic species, arsenite [As^{3+}] and arsenate [As^{5+}], are the predominant species in most environments, although the organic ones might also be present [15].

In general, inorganic compounds of arsenic are regarded as more highly toxic than most organic forms which are less toxic [10, 14, 16, 17]. The trivalent compounds (arsenites) are more toxic than the pentavalent compounds (arsenates) [16, 17]. It has been reported that As^{3+} is 4 to 10 times more soluble in water than As^{5+}. However, the trivalent methylated arsenic species have been found to be more toxic than inorganic arsenic because they are more efficient at causing DNA breakdown [17]. Although As^{5+} tends to be less toxic compared to of As^{3+}, it is thermodynamically more stable due to it predominates under normal conditions and becomes the cause of major contaminant in ground water [14]. Arsenate which is in the pentavalent state (As^{5+}) is also considered to be toxic and carcinogenic to human [18].

Lead (Pb)

Lead (Pb), with atomic number 82, atomic weight 207.19, and a specific gravity of 11.34, is a bluish or silvery-grey metal with a melting point of 327.5°C and a boiling point at atmospheric pressure of 1740°C. It has four naturally occurring isotopes with atomic weights 208, 206, 207 and 204 (in decreasing order of abundance). Despite the fact that lead has four electrons on its valence shell, its typical oxidation state is +2 rather than +4, since only two of the four electrons ionize easily. Apart from nitrate, chlorate, and chloride, most of the inorganic salts of $lead^{2+}$ have poor solubility in water [19]. Lead (Pb) exists in many forms in the natural sources throughout the world and is now one of the most widely and evenly distributed trace metals. Soil and plants can be contaminated by lead from car exhaust, dust, and gases from various industrial sources. Pb^{2+} was found to be acute toxic to human beings when present in high amounts. Since Pb^{2+} is not biodegradable, once soil has become contaminated, it remains a long-term source of Pb^{2+} exposure. Metal pollution has a harmful effect on biological systems and does not undergo biodegradation [7].

Soil can be contaminated with Pb from several other sources such as industrial sites, from leaded fuels, old lead plumbing pipes, or even old orchard sites in production where lead arsenate is used. Lead accumulates in the upper 8 inches of the soil and is highly immobile. Contamination is long-term. Without remedial action, high soil lead levels will never return to normal [20].

In the environment, lead is known to be toxic to plants, animals, and microorganisms. Effects are generally limited to especially contaminated areas [21]. Pb contamination in the environment exists as an insoluble form, and the toxic metals pose serious human health problem, namely, brain damage and retardation [5].

Mercury (Hg)

Mercury is a naturally occurring metal that is present in several forms. Metallic mercury is shiny, silver-white, odorless liquid. Mercury combines with other elements, such as chlorine, sulfur, or oxygen, to form inorganic mercury compounds or salts, which are usually white powders or crystals. Mercury also combines with carbon to make organic mercury compounds [22]. Mercury, which has the lowest melting point (−39°C) of all the pure metals, is the only pure metal that is liquid at room temperature. However, due to its several physical and chemical advantages such as its low boiling point (357°C) and easy vaporization, mercury is still an important material in many industrial products [23]. As any other metal, mercury could occur in the soil in various forms. It dissolves as free ion or soluble complex and is nonspecifically adsorbed by binding mainly due to the electrostatic forces, chelated, and precipitated as sulphide, carbonate, hydroxide, and phosphate. There are three soluble forms of Hg in the soil environment. The most reduced is

Hg^0 metal with the other two forms being ionic of mercurous ion Hg_2^{2+} and mercuric ion Hg^{2+}, in oxidizing conditions especially at low pH. Hg^+ion is not stable under environmental conditions since it dismutates into Hg^0 and Hg^{2+}. A second potential route for the conversion of mercury in the soil is methylation to methyl or dimethyl mercury by anaerobic bacteria [24].

Mercury is a persistent environmental pollutant with bioaccumulation ability in fish, animals, and human beings [23]. Mercury salts and organomercury compounds are among the most poisonous substances in our environment. The mechanism and extent of toxicity depend strongly on the type of compound and the redox state of mercury [25].

Environmental contamination due to mercury is caused by several industries, petrochemicals, minings, painting, and also by agricultural sources such as fertilizer and fungicidal sprays [26]. Some of the more common sources of mercury found throughout the environment include but may not be limited to the household bleach, acid, and caustic chemicals (e.g., battery acid, household lye, muriatic acid (hydrochloric acid), sodium hydroxide, and sulfuric acid), instrumentation containing mercury (e.g., medical instruments, thermometers, barometers, and manometers), dental amalgam (fillings), latex paint (manufactured prior to 1990), batteries, electric lighting (fluorescent lamps, incandescent wire filaments, mercury vapor lamps, ultraviolet lamps), pesticides, pharmaceuticals (e.g., nasal sprays, cosmetics, contact lens products), household detergents and cleaners, laboratory chemicals, inks and paper coatings, lubrication oils, wiring devices and switches, and textiles. Though mercury use in many of the above items being produced now is restricted or banned, there are still some existing, older products in use [22].

Terrestrial plants are generally insensitive to the harmful effects of mercury compounds; however, mercury is known to affect photosynthesis and oxidative metabolism by interfering with electron transport in chloroplasts and mitochondria. Mercury also inhibits the activity of aquaporins and reduces plant water uptake [27].

Mercury and its compounds are cumulative toxins and in small quantities are hazardous to human health. The major effects of mercury poisoning manifest as neurological and renal disturbances as it can easily pass the blood-brain barrier and has effect on the brain [26].

PHYTOREMEDIATION TECHNOLOGY

Phytoremediation techniques have been briefly depicted in many literatures or articles. The generic term "phytoremediation" consists of the Greek prefix phyto (plant), attached to the Latin root remedium (to correct or remove an evil) [28, 29]. Some definitions on phytoremediation that have been described by several researchers are listed in Table 1.

Table 1: Definition of phytoremediation

No.	Researchers	Definition of phytoremediation
(1)	[30]	The use of plants to improve degraded environments
(2)	[31]	The use of plants, including trees and grasses, to remove, destroy or sequester hazardous contaminants from media such as air, water, and soil
(3)	[24]	The use of plants to remediate toxic chemicals found in contaminated soil, sludge, sediment, ground water, surface water, and wastewater
(4)	[32]	An emerging technology using specially selected and engineered metal accumulating plants for environmental cleanup
(5)	[33]	The use of vascular plants to remove pollutants from the environment or to render them harmless

(6)	[3]	The engineered use of green plant to remove, contain, or render harmless such environmental contaminants as heavy metals, trace elements, organic compounds, and radioactive compounds in soil or water. This definition includes all plant-influenced biological, chemical, and physical processes that aid in the uptake, sequestration, degradation, and metabolism of contaminants, either by plants or by the free-living organisms that constitute the plant rhizosphere
(7)	[29]	Phytoremediation is the name given to a set of technologies that use different plants as a containment, destruction, or an extraction technique. Phytoremediation is an emerging technology that uses various plants to degrade, extract, contain, or immobilize contaminants from soil and water
(8)	[34]	Phytoremediation in general implies the use of plants (in combination with their associated microorganisms) to remove, degrade, or stabilize contaminants

Generally, according to the above researchers, phytoremediation is defined as an emerging technology using selected plants to clean up the contaminated environment from hazardous contaminant to improve the environment quality. Figure 1 depicts the uptake mechanisms of both organics and inorganics contaminants through phytoremediation technology. For organics, it involves phytostabilization, rhizodegradation, rhizofiltration, phytodegradation, and phytovolatilization. These mechanisms related to organic contaminant property are not able to be absorbed into the plant tissue. For inorganics, mechanisms which can be involved are phytostabilization, rhizofiltration, phytoaccumulation and phytovolatilization.

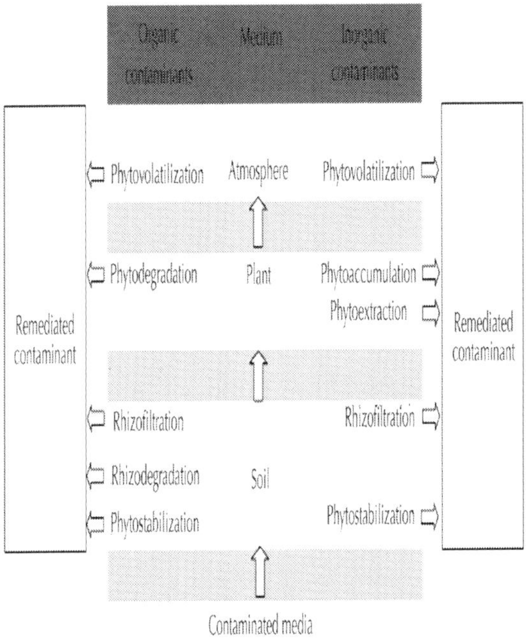

Figure 1: Uptake mechanisms on phytoremediation technology. Source: [35].

Based on Figure 1, some certain essential processes involved in phytoremediation technology [29, 31] are phytostabilization and phytoextraction for inorganic contaminants, and phytotransformation/phytodegradation, rhizofiltration, and rhizodegradation for organic contaminants.

The root plants exudates to stabilize, demobilize and bind the contaminants in the soil matrix, thereby reducing their bioavailability. These all are called as phytostabilization process. Certain plant species have used to immobilize contaminants in the soil and ground water through absorption and accumulation by roots, adsorption onto roots, or precipitation within the root zone. This process is for organics and metals contaminants in soils, sediments, and sludges medium [29, 31].

Specific plant species can absorb and hyperaccumulate metal contaminants and/or excess nutrients in harvestable root and shoot tissue, from the growth substrate through phytoextraction process. This is for metals, metalloids, radionuclides, nonmetals, and organics contaminants in soils, sediments, and sludges medium [29, 31]. Phytovolatilization process is the plants ability to absorb and subsequently volatilize the contaminant into the atmosphere. This process is for metal contaminants in groundwater, soils, sediments, and sludges medium. Since phytotransformation/phytodegradation process is the breakdown of contaminants taken up by plants through metabolic processes within the plant or the breakdown of contaminants externally to the plant through the effect of compounds produced by the plants. This process is for complex organic molecules that are degraded into simpler molecule contaminants in soils, sediments, sludges, and groundwater medium [29,31].

Plant roots take up metal contaminants and/or excess nutrients from growth substrates through rhizofiltration (=root) process, the adsorption, or, precipitation onto plant roots or absorption into the roots of contaminants that are in solution surrounding the root zone. This process is for metals, excess nutrients, and radionuclide contaminants in groundwater, surface water, and wastewater medium [29, 31].

The breakdown of contaminants in the soil through microbial activity that is enhanced by the presence of the root zone is called rhizodegradation. This process uses microorganisms to consume and digest organic substances for nutrition and energy. Natural substances released by the plant roots, sugars, alcohols, and acids, contain organic carbon that provides food for soil microorganisms and establish a dense root mass that takes up large quantities of water. This process is for organic substance contaminants in soil medium [29, 31].

MECHANISMS OF HEAVY METAL UPTAKE BY PLANT

Contaminant uptake by plants and its mechanisms have been being explored by several researchers. It could be used to optimize the factors to improve the performance of plant uptake. According to Sinha et al. [36], the plants act both as "accumulators" and "excluders". Accumulators survive despite concentrating contaminants in their aerial tissues. They biodegrade or biotransform the contaminants into inert forms in their tissues. The excluders restrict contaminant uptake into their biomass.

Plants have evolved highly specific and very efficient mechanisms to obtain essential micronutrients from the environment, even when present at low ppm levels. Plant roots, aided by plant-produced chelating agents and plant-induced pH changes and redox reactions, are able to solubilize and take up micronutrients from very low levels in the soil, even from nearly insoluble precipitates. Plants have also evolved highly specific mechanisms to translocate and store micronutrients. These same mechanisms are also involved in the uptake, translocation, and storage of toxic elements, whose chemical properties simulate those of essential elements. Thus, micronutrient uptake mechanisms are of great interest to phytoremediation [37].

The range of known transport mechanisms or specialized proteins embedded in the plant cell plasma membrane involved in ion uptake and translocation include (1) proton pumps ("-ATPases that consume energy and generate electrochemical gradients), (2) co- and antitransporters (proteins that use the electrochemical gradients generated by "-ATPases to drive the active uptake of ions), and (3) channels (proteins that facilitate the transport of ions into the cell). Each transport mechanism is likely to take up a range of ions. A basic problem is the interaction of ionic species during uptake of various heavy metal contaminants. After uptake by roots, translocation into shoots is desirable because the harvest of root biomass is generally not feasible. Little is known regarding the forms in which metal ions are transported from the roots to the shoots [37].

Plant uptake-translocation mechanisms are likely to be closely regulated. Plants generally do not accumulate trace elements beyond near-term metabolic needs. And these requirements are small ranging from 10 to 15 ppm of most trace elements suffice for most needs [37]. The exceptions are "hyperaccumulator" plants, which can take up toxic metal ions at levels in the thousands of ppm. Another issue is the form in which toxic metal ions are stored in plants, particularly in hyperaccumulating plants, and how these plants avoid metal toxicity. Multiple mechanisms are involved. Storage in the vacuole appears to be a major one [37].

Water, evaporating from plant leaves, serves as a pump to absorb nutrients and other soil substances into plant roots. This process, termed evapotranspiration, is responsible for moving contamination into the plant shoots as well. Since contamination is translocated from roots to the shoots, which are harvested, contamination is removed while leaving the original soil undisturbed. Some plants that are used in phytoextraction strategies are termed "hyperaccumulators." They are plants that achieve a shoot-to-root metal-concentration ratio greater than one. Nonaccumulating plants typically have a shoot-to-root ratio considerably less than one. Ideally, hyperaccumulators should thrive in toxic environments, require little maintenance and produce high biomass, although few plants perfectly fulfill these requirements [38].

Metal accumulating plant species can concentrate heavy metals like Cd, Zn, Co, Mn, Ni, and Pb up to 100 or 1000 times those taken up by nonaccumulator (excluder) plants. In most cases, microorganisms bacteria and fungi, living in the rhizosphere closely associated with plants, may contribute to mobilize metal ions, increasing the bioavailable fraction. Their role in eliminating organic contaminants is even more significant than that in case of inorganic compounds [39, 40].

Heavy metal uptake by plant through phytoremediation technologies is using these mechanisms of phytoextraction, phytostabilisation, rhizofiltration, and phytovolatilization as shown in Figure 2.

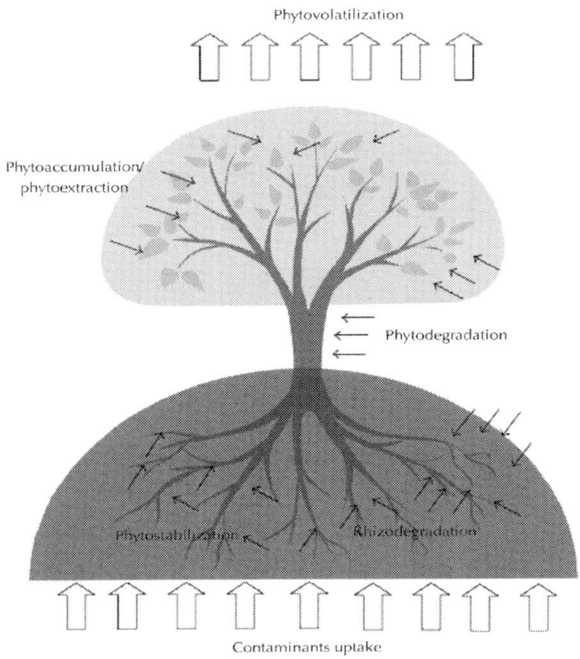

Figure 2: The mechanisms of heavy metals uptake by plant through phytoremediation technology.

Phytoextraction

Phytoextraction is the uptake/absorption and translocation of contaminants by plant roots into the above ground portions of the plants (shoots) that can be harvested and burned gaining energy and recycling the metal from the ash [28, 39–42].

Phytostabilisation

Phytostabilisation is the use of certain plant species to immobilize the contaminants in the soil and groundwater through absorption and accumulation in plant tissues, adsorption onto roots, or precipitation within the root zone preventing their migration in soil, as well as their movement by erosion and deflation [28,39–42].

Rhizofiltration

Rhizofiltration is the adsorption or precipitation onto plant roots or absorption into and sequesterization in the roots of contaminants that are in solution surrounding the root zone by constructed wetland for cleaning up communal wastewater [28, 39–42].

Phytovolatilization

Phytovolatilization is the uptake and transpiration of a contaminant by a plant, with release of the contaminant or a modified form of the contaminant to the atmosphere from the plant. Phytovolatilization occurs as growing trees and other plants take up water along with the contaminants. Some of these contaminants can pass through the plants to the leaves and volatilize into the atmosphere at comparatively low concentrations [28, 39–42].

Plants also perform an important secondary role in physically stabilizing the soil with their root system, preventing erosion, protecting the soil surface, and reducing the impact of rain. At the same time, plant roots release nutrients that sustain a rich microbial community in the rhizosphere. Bacterial community composition in the rhizosphere is affected by complex interactions between soil type, plant species, and root zone location. Microbial populations are generally higher in the rhizosphere than in the root-free soil. This is due to a symbiotic relationship between soil microorganisms and plants. This symbiotic relationship can enhance some bioremediation processes. Plant roots also may provide surfaces for sorption or precipitation of metal contaminants [27].

In phytoremediation, the root zone is of special interest. The contaminants can be absorbed by the root to be subsequently stored or metabolised by the plant. Degradation of contaminants in the soil by plant enzymes exuded from the roots is another phytoremediation mechanism [43]. For many contaminants, passive uptake via micropores in the root cell walls may be a major route into the root, where degradation can take place [3].

FACTORS AFFECTING THE UPTAKE MECHANISMS

There are several factors which can affect the uptake mechanism of heavy metals, as shown in Figure 3. By having knowledge about these factors, the uptake performance by plant can be greatly improved.

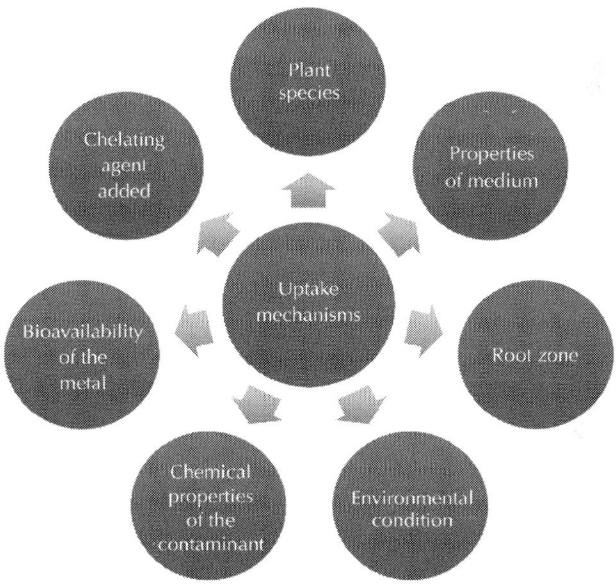

Figure 3: Factors which are affecting the uptake mechanisms of heavy metals.

The Plant Species

Plants species or varieties are screened, and those with superior remediation properties are selected [31]. The uptake of a compound is affected by plant species characteristic [44]. The success of the phytoextraction technique depends upon the identification of suitable plant species that hyperaccumulate heavy metals and produce large amounts of biomass using established crop production and management practices [24].

Properties of Medium

Agronomical practices are developed to enhance remediation (pH adjustment, addition of chelators, fertilizers) [31]. For example, the amount of lead absorbed by plants is affected by the pH, organic matter, and the phosphorus content of the soil. To reduce lead uptake by plants, the pH of the soil is adjusted with lime to a level of 6.5 to 7.0 [20].

The Root Zone

The Root Zone is of special interest in phytoremediation. It can absorb contaminants and store or metabolize it inside the plant tissue. Degradation of contaminants in the soil by plant enzymes exuded from the roots is another phytoremediation mechanism. A morphological adaptation to drought stress is an increase in root diameter and reduced root elongation as a response to less permeability of the dried soil [43].

Vegetative Uptake

Vegetative Uptake is affected by the environmental conditions [44]. The temperature affects growth substances and consequently root length. Root structure under field conditions differs from that under greenhouse condition [43]. The success of phytoremediation, more specifically phytoextraction, depends on a contaminant-specific hyperaccumulator [45]. Understanding mass balance analyses and the metabolic fate of pollutants in plants are the keys to proving the applicability of phytoremediation [46].

Metal uptake by plants depends on the bioavailability of the metal in the water phase, which in turn depends on the retention time of the metal, as well as the interaction with other elements and substances in the water. Furthermore, when metals have been bound to the soil, the pH, redox potential, and organic matter content will all affect the tendency of the metal to exist in ionic and plant-available form. Plants will affect the soil through their ability to lower the pH and oxygenate the sediment, which affects the availability of the metals [47], increasing the bioavailability of heavy metals by the addition of biodegradable physicochemical factors, such as chelating agents and micronutrients [34].

Addition of Chelating Agent

The increase of the uptake of heavy metals by the energy crops can be influenced by increasing the bioavailability of heavy metals through addition of biodegradable physicochemical factors such as chelating agents, and micronutrients, and also by stimulating the heavy-metal-uptake capacity of the microbial community in and around the plant. This faster uptake of heavy metals will result in shorter and, therefore, less expensive remediation periods. However, with the use of synthetic chelating agents, the risk of increased leaching must be taken into account [34]. The use of chelating agents in heavy-metal-contaminated soils could promote leaching of the contaminants into the soil. Since the bioavailability of heavy metals in soils decreases above pH 5.5–6, the use of a chelating agent is warranted, and may be required, in alkaline soils. It was found that exposing plants to EDTA for a longer period (2 weeks) could improve metal translocation in plant tissue as well as the overall phytoextraction performance. The application of a synthetic chelating agent (EDTA) at 5 mmol/kg yielded positive results [8]. Plant roots exude organic acids such as citrate and oxalate, which affect the bioavailability of metals. In chelate-assisted phytoremediation, synthetic chelating agents such as NTA and EDTA are added to enhance the phytoextraction of soil-polluting heavy metals. The presence of a ligand affects the biouptake of heavy metals through the formation of metal-ligand complexes and changes the potential to leach metals below the root zone [48].

EFFECTIVENESS OF HEAVY METALS UPTAKEBY PLANTS

Several studies have described the performance of heavy metals uptake by plants. It is reported that phytoremediation technology is an alternative to treat heavy-metal-contaminated side which will be more admitted in order to remediate the environment. Table 2 lists some research done to remediate heavy metals from contaminated soil, while Table 3 lists some research conducted to remediate them from contaminated water and wastewater.

Table 2: Phytoremediation study on soil medium

No.	Researcher	Research scale and duration	Uptake mechanisms and media (substrate)	Contaminant or parameter and concentration	Plants name and type	Result
(1)	[16]	Greenhouse pot experiment (6, 10, and 16 days)	Phytoextraction (soil was added to aqueous solution and was dried overnight in an oven at 120°C, cooled, and transferred to the pot)	Aqueous solution containing 0.1041 g of sodium arsenate heptahydrate ($Na_2HAsO_4 \cdot 7H_2O$), the mixture which contained 50 mg/kg of As (wet weight)	Leersia oryzoides(rice-cut grass)—terrestrial plant	The increase in plant size is matched by a decrease in shoot arsenic concentration. The data show that 12, 13, and 13 mg/m^{2}of arsenic were absorbed by the shoots at 6, 10, and 16 weeks, respectively. Since the SRQ and PECs all exhibit the same downward trend after 6 wk, it is suggested that periodic mowing ofLeersia oryzoides grown for phytoextraction purposes on contaminated land could maintain the high arsenic uptake at 6 week.
(2)	[33]	Laboratory (pot experiment) (90 days)	Fly ash and soil mixtures	Pb as lead nitrate, Zn as zinc sulfate, Ni as nickel sulfate, Mn as manganese chloride, and Cu as copper sulfate (1000 ppm concentration each (Spiked))	Scirpus littoralis—semiaquatic	The metal content ratios BO/soil (B/S) were higher than shoot/soil ratios (T/S) for all the metals, the highest being for Ni. Metal ratios BO/water (B/W) were also higher than shoot/water (T/W) ratios, but the B/W ratio was maximum for Zn. All the metals except Ni showed negative correlation with nitrogen but they were all nonsignificant. However, Puptake showed positive correlations with all the metals, and all were significant at 1% confidence limit.

(3)	[49]	Field study (90 days)	Soil (agricultural land area)	(Cu, Cd, Cr, Zn, Fe, Ni, Mn, and Pb)	Wheat (Triticum aestivum L.)—terrestrial Indian mustard (Brassica campestris L.)—terrestrial	Analyses of effluents and soil samples have shown high metal content than the permissible limit except Pb. Analyses of plant samples have indicated the maximum accumulation of Fe followed by Mn and Zn in root > shoot > leaves > seeds. Maximum increase in photosynthetic pigment was observed between 30 and 60 days while protein content was found maximum between 60 and 90 days of growth period in both plants.
(4)	[5]	Laboratory (65 days)	Phytoextraction (soil)	Pb by using standard Pb solutions (75 mg Pb/1 kg soil)	Creeping zinnia (Alternanthera phyloxeroides)—aquatic Moss rose (Sanvitalia procumbens)—terrestrial Alligator weed (Portulaca grandiflora)—aquatic	Alternanthera phyloxeroides shows the highest lead content in its tissues. This might be caused by it forming long stolons, a massive fibrous root system, and large surface area which benefits the accumulation of lead. Efficiency process 30–80%.
(5)	[34]	Literature review	Soil	Cd, Cr, Cu, Ni, Pb, and Zn	Brassica juncea(Indian mustard),Brassica rapa (field mustard), andBrassica napus(rape)—terrestrial	Brassica rapa exhibited the highest affinity for accumulating Cd and Pb from the soil, either with/without additional use of mobilizing soil amendments. Two Brassica species (Brassica napus and Raphanus sativus) were moderately tolerant when grown on a multi-metalcontaminated soil. The distribution of heavy metals in the organs of crops decreased in the following order: leaves > stems > roots > fruit shell > seeds.
(6)	[50]	Laboratory—pot experiment (12 days)	Agropeat and half strength Hoagland solution	Arsenic (As) as of sodium (meta-) arsenite (50 uM, 150 uM and 300 uM)	Brassica juncea var. Varuna and Pusa Bold—terrestrial	Increase/decrease of antioxidant enzymes activities showed not much changes at the given concentrations. The data presented indicates the differential responses in both the varieties and also that the increased tolerance in P. Bold may be due to the defensive role of antioxidant enzymes, induction of MAPK, and upregulation of PCS transcript which is responsible for the production of metal-binding peptides.

(7)	[51]	Field study (trials to extract heavy metals from two contaminated soils, one calcareous (5 years) and one acidic (2 years))	Phytoextraction (soil)	Cd and Zn	Willow (Salix viminalis)—terrestrial	Salix had performed better on the acidic soil because of larger biomass production and higher metal concentrations in shoots. Addition of elemental sulphur to the soil did not yield any additional benefit in the long term, but application of an Fe chelate improved the biomass production. Cd and Zn concentrations were significantly higher in leaves than stems. On both soils, concentration in shoots decreased with time.
(8)	[52]	Laboratory (26 days)	Sludge-amended soils	Cd and Zn	Raphanus sativus L.	This study has shown that clear evidence of as ludge-driven plateau response in metal uptake by plants will only be obtained when studies have found a good hyperbolic relationship between soil solution metal concentration with increasing sludge application rate and can link this to a plateau response in plant uptake of metals.
		Laboratory—lysimeter pot (March 1995–September 1995)	Soil	Zn as $ZnSO_4$(50, 1,500, 2,000 µg/g (ppm) Zn. and 2,000 µg/g (ppm), and 0 µg/g (ppm) (control) received nutrient only)	Hybrid poplar (Populus sp.)—terrestrial	At levels of zinc above 1,000 µg/g (ppm) in nutrient added, leachate levels were always below 100 µg/g (ppm) in samples as the zinc addition; these levels increased the following day and then decreased sharply the second day after the zinc addition, to concentrations less than 100 µg/g (ppm). The zinc concentration steadily decreased as the plants apparently reabsorbed the zinc as the nutrient was cycled through the pots on subsequent days. The root tissues showed much higher concentrations of accumulated and sequestered metal than did the above ground parts.

(9)	[3]	Laboratory (April 1996, 2 months)	Soil	Zn (160 μg/g Zn, 600 μg/g Zn, and 0 μg/g Zn (control))	Eastern gamagrass (Tripsacum dactyloides)—terrestrial	Leachate analyses for zinc indicate that initially plants subjected to both levels of zinc were removing up to 70% of the zinc from the leachate. The plants receiving 160 μg/g Zn had grown considerably and were almost the same size as the controls (no zinc), but some of the mature leaf blades were rolled; the mean zinc removal rate for these plants was 50% of the zinc in the leachate. The plants receiving 600 μg/g Zn were smaller than the controls, their color was a darker green, most of the mature leaf blades were rolled, and the mean zinc removal rate was about 30% of the zinc in the leachate.
			Soil	Pb and As (up to 1000 μg/g Pb and up to 200 μg/g As)	Hybrid willow (Salix sp.) and hybrid poplar (Populus sp.)—terrestrial	The willows were able to remove approximately 9.5% of the available lead and about 1% of the total arsenic from the contaminated soil. The less mature poplars removed about 1% of the available lead and 0.1% of the total arsenic from the same soil. In the sand experiment, the willows took up about 40% of the administered lead and 30 to 40% of the administered arsenic.
(10)	[53]	Field (1976–2001)	(Soil)	Nonessential (Cd, Ni, Pb) and essential heavy metals (Cu, Fe, Mn, Zn). The tetrasodium salt of EDTA was applied at rates of 0, 0.5, 1, 2 g EDTA salt/kg surface (25 cm depth) soil	Sunflower (Helianthus annuusL.) and Hybrid poplar (Populus deltoides Marsh. x P. nigra L.)—terrestrial	For sunflower, the 1.0 g/kg rate of chelate addition resulted in maximal removal of the three nonessential heavy metals (Cd, Ni, Pb). Uptake of the essential heavy metals by sunflower was little affected by the EDTA. The leaves of sunflower grown with 1.0 g EDTA $Na_4 \cdot 2H_2O$/kg soil accumulated more Cd, Ni, and Pb than leaves of sunflower grown without the EDTA salt. Removal of the non-essential heavy metals by sunflower was greater at the higher plant density compared to the lower one.

(11)	[54]	Laboratory	18 different phytoremedation treatments. I. parcel: mine waste without fly ash. Control and untreated plot. 3 test plants. II. mine waste + fly ash without liming. Control and untreated plot. 3 test plants. III. mine waste + fly ash + liming. Control and untreated plot. 3 test plants.	As, Cd, Mo, Pb, Zn (soil) and As, Cd, Pb, Ni, Zn (water)	Grasses (mixture of selected species), sorghum (Sorghum bicolor L.) and Sudan grass (Sorghum sudanense)—terrestrial	The chemical risks of the Gyöngyösoroszi spoils were assessed. The major contaminants of the waste mine were identified: Pb, Zn, Cd, As. The concept of the integrated phytoremediation was successfully applied to vegetate Gyöngyösoroszi spoil. The biomass production was different, depending on the technology variant. The highest biomass production was achieved, when multilevel revitalization was also applied. The integrated phytoremediation treatments not only produced high biomass, but also decreased the heavy metal content in the plants.
(12)	[55]	Field (1995–1997)	Soil	Ni, Cu, Cd, Zn	Willow (Salix spp.)—terrestrial	One group of willow had relatively low Ni and Cu in the bark and high Cd and Zn in the wood, with a good survival rate and biomass production. The second group of willow had relatively high Ni and Cu in the bark and low Cd and Zn in the wood and performed poorly in terms of survival and biomass production.
(13)	[8]	Laboratory (15 May and 25 September 2002)	Phytoextraction (soil)	Cu, Pb, Zn	Fescue (Festuca arundinaceaSchreb.), Indian mustard (Brassica juncea (L.) Czern.), and willow (Salix viminalis L.)—terrestrial	The use of the freeacid form of EDTA and exposure time of one to two weeks before harvesting increased the concentration of metals translocated to plant tissues. It is found no significant difference in heavy metal concentrations in higher and lower soil horizons between EDTA treated and untreated soils. Exposing plants to EDTA for a longer period (2 weeks) could improve metal translocation in plant tissue as well as the overall phytoextraction performance.

(14)	[24]	Field experiment (3 years)	Phytoextraction (soil content with Hg)	Hg (mean Hg content of the soil was 29.17 μg/g for the 0–10 cm horizon and 20.32 μg/g for 10–40 cm horizon with less than 2% of the total Hg being bioavailable)	Three agriculture crop plants:Triticum aestivum(wheat)—terrestrial Hordeum vulgare(barley)—terrestrial Lupinus luteus(yellow lupin)—terrestrial	The decrease of mean Hg concentration from 29.17 μg g–1 at 0–10 cm horizon to 20.32 μg g–1 at 10–40 cm horizon demonstrated the anthropogenic origin of the mercury in the soil. Preliminary results show that all crops extracted mercury, with Hg plant concentration reaching up to 0.479 μg g–1 in wheat. The mercury concentration in the plants accounted for less than 3% of mercury concentration in the soil. The Hg concentrations in the plants were similar or even higher than that of the bioavailable Hg in the soils. Mercury extraction yields reached up to 719 mg/ha for barley.
(15)	[56]	Pot experiment (20 weeks)	Soil from waste deposits of the lead smelter	Pb	Agrostis capillaris—terrestrial	Inoculation with indigenous or nonindigenous AMF in this experiment did not decrease Pb uptake by the host in comparison with nonmycorrhizal plants grown in contaminated soil. It can be concluded that 13 months of subculturing in an inert substrate did not affect development of G. intraradicesPH5 isolated from the waste deposits of a Pb smelter in contaminated soil of its origin. The interaction of the fungus with the host plant was changed: the ability of the lineage cultured without HM to support plant growth in Pb-contaminated soil was decreased, while translocation of Pb from plant roots to shoots increased.
(16)	[38]	Field and greenhouse experiments	Phytoextraction (As- and Pb-contaminated soil)	Arsenic (As) and lead (Pb)	Chinese Brake Ferns (Pteris vittata)—terrestrial Indian Mustard (Brassica juncea)—terrestrial	It appears that EDTA is necessary for Pb extraction due to the low soil Pb bioavailability. Soil amendments like EDTA are necessary because they mobilize soil Pb, making it available to plant roots. It may not be advisable to apply EDTA in the environment, because EDTA mobilizes metals, which may leach into surrounding property of groundwater. The presence of other metals that compete for EDTA may increase the amount of EDTA required for Pb remediation.

(17)	[27]	Laboratory experiment, using chamber (6 weeks)	Phytostabilization (mercury-contaminated soil used in this experiments was obtained from a chemical factory located in the southeast part of Poland, which has been in operation for over 50 years)	Hg	Species Festuca rubra (red fescue)—terrestrial Poa pratensis (meadow grass)—terrestrial Armoracia lapathifolia (horseradish)—terrestrial Helianthus tuberosus (Jerusalem sunflower)—terrestrial S. viminalis (willow)— terrestrial	The highest concentrations of mercury were found at the roots, but translocation to the aerial part also occurred. Most of the plant species tested displayed good growth on mercury contaminated soil and sustained a rich microbial population in the rhizosphere. An inverse correlation between the number of sulfur amino acid decomposing bacteria, and root mercury content was observed. These results indicate the potential for using some species of plants to treat mercury-contaminated soil through stabilization rather than extraction.
(18)	[57]	Field (July and October)	Phytoextraction and phytostabilisation (soil)	Zn, Cu, Cr and Cd	Two poplar clones (Populus deltoides x maximowiczii-clone Eridano andP. x euramericana-clone I-214)—terrestrial	Leaf, stem, root and woody cutting biomasses of treated plants were significantly greater than those in the controls in both clones, except for stem biomass at the beginning of October. Among the four heavy metals (Zn, Cu, Cr, and Cd), only Zn, Cu, and Cr concentrations in plants differed consistently between clones or soil treatments, while Cd levels were always below the detection limits.
(19)	[58]	Field study and laboratory experiment (2002-2003 (field study), 3 months for laboratory experiment)	Soil	Fe, Zn, Pb, Cu, Ni, Cr, Mn	Brachythecium populeum	The results obtained from this study onB. populeum lead to the inference that physiological/bio-chemical analysis of epiphytic bryophytes can serve as cost-effective indicators/monitors for the environmental quality of any area, and on the basis of this information appropriate steps can be taken to improve the air quality of an area.

(20)	[59]	Pot experiment and field trial (2004-2005 for pot experiment, and 2005 field trial)	Phytoextraction and phytostabilization (soil)	As, Co, Cu, Pb, and Zn	Three poplar species (Populus alba, Populus nigra, Populus tremula) and Salix alba—terrestrial	Trace element concentrations were much higher in roots than in above-ground tissues, with particularly high concentrations in fine roots. The highest accumulations were measured in P. nigraand S. alba. In wood, the highest concentrations of Cu and Zn were in S. alba. Salix alba foliage contained highest concentrations of As, Cu, Pb, and Zn; leaf Zn concentration exceeded those of wood by almost 6 times. The overall removal of trace elements was only significantly higher in P. alba than in S. alba; P. alba.
(21)	[60]	Pot experiment and field trial (2 years (2004-2005) for pot experiment and field trial on May–September 2005)	Phytoextraction and phytostabilization (soil (Pyrite ore contains mainly pyrite (FeS_2), lesser amounts of chalcopyrite ($CuFeS_2$), sphalerite (ZnS), magnetite (Fe_3O_4), and various trace elements))	As, Co, Cu, Pb and Zn	P. alba L. (white poplar)—terrestrial P. nigra L. (black poplar)—terrestrial P. tremula L. (European aspen)—terrestrial Salix alba L. (white willow)—terrestrial	The result shown that establishment ofPopulus and Salix species at the site is achievable through ripping of the surface, minimal tillage, some mixing of the wastes with imported soil, irrigation and fertilisers. Potentially, the elevated concentrations of Pb, As and other elements could be leached from the remediated wastes towards groundwater or other receptors, and these fluxes could also be influenced by soil amendments, changes in the rhizosphere or both. Immobilisation of trace elements in both coarse and fine roots may reduce leaching, particularly of Cu and Zn but also As and Pb.
(22)	[61]	Greenhouse	Phytoextraction and phytostabilization (soil)	Six sediment-derived soils with increasing field Cd levels (0.9–41.4 mg/kg)	Two willow clones (Salix fragilis "Belgisch Rood" andSalix viminalis "Aage")—terrestrial	No growth inhibition was observed for both clones for any of the treatments. Dry weight root biomass and total shoot length were significantly lower for S. viminalis compared to S. fragilis for all treatments. Willow foliar Cd concentrations were strongly correlated with soil and soil water Cd concentrations. Both clones exhibited high accumulation levels of Cd and Zn in above ground plant parts. Cu, Cr, Pb, Fe, Mn, and Ni were found mainly in the roots. Bioconcentration factors of Cd and Zn in the leaves were the highest for the treatments with the lowest soil Cd and Zn concentration.

(23)	[62]	Laboratory and field	Rhizobox experiment was used to investigate the short-term effect of willow roots on metal availability in oxic and anoxic sediment. Longer-term effects were assessed in a field trial (soil)	Cd, Zn, Cu, and Pb	Willow (Salix spp.)—terrestrial	The rhizobox trial showed that Cd, Zn, and Cu extractability in the rhizosphere increased while the opposite was observed for Pb. The field trial showed that Cu and Pb, but not Cd, were more available in the root zone after water and ammonium acetate (pH 7) extraction compared with the bulk sediment. Sediment in the root zone was better structured and aggregated and thus more permeable for downward water flows, causing leaching of a fraction of the metals and significantly lower total contents of Cd, Cu, and Pb.
(24)	[63]	Pot experiment	Phytoextraction (soil)	As (as Na_2HAsO_4), Cd (as $CdCl_2$), Pb (as $Pb(CH_3COO)_2$), and Zn (as $Zn(CH_3COO)_2$) (100 mg As/kg, 40 mg Cd/kg, 2000 mg Pb/kg, and 2000 mg Zn/kg)	Salix spp.—terrestrial	Although As and Cd uptake slightly increased in Suchdol-Zn soil compared to Suchdol-Pb soil, the element removal from soil was significantly higher in Suchdol-Pb soil due to a significant reduction of aboveground biomass yield in Suchdol-Zn soil. The yield reduction decreased the uptake of plant-available elements by biomass; thus higher plant-available portions of As and Cd were found in Suchdol-Zn soil.
(25)	[64]	Field survey: from 12 As-contaminated sites (September to November 2003)	Field study: contaminated soil	As	Samples of 24 fern species belonging to 16 genera and 11 families as well as their associated soils were collected—terrestrial	Pteris multifida and P. oshimensis can (hyper-) accumulate As in their fronds with high concentrations. Total As concentrations in soils associated with P. multifida and P. oshimensis varied from 1262 to 47,235 mg/kg, but the DTPA-extractable As concentrations were relatively low, with a maximum of 65 mg/kg. Although As concentrations in the fronds of P. oshimensis were comparatively lower than those of P. multifida, its high above ground biomass makes it more suitable for phytoremediating As-contaminated soils.

(26)	[65]	Field survey (contaminated site since 1976; the sample was taken in 2006)	Soil	Cu, Pb, Cd, and Zn	Paulowni fortunei(seem) Hems	In the rhizosphere and bulk soils of P. fortunei, all physico-chemical properties increased with the revegetation time. The total contents of Cu, Pb, Cd, and Zn also consistently increased with the re-vegetation time; moreover, rhizosphere soils accumulated more heavy metals than bulk soils with the revegetation time. In the rhizosphere soils of P. fortunel, the immobility and bioavailability of heavy metals were enhanced. In the rhizosphere microenvironment, pH, OM, and EC were important factors affecting the distribution of heavy metal fractions. Among different heavy metal fractions, the exchangeable and organically bound fractions were easily available for P. fortunei, but carbonate, Fe–Mn oxide, and residual fractions were not easily available for P. fortunei.
(27)	[4]	Greenhouse pot experiment (August-September 2002)	Soil was sampled in two sites: contaminated soil was taken near road with heavy traffic and clean soil was taken from park protected from the road by buildings	Ag, As, B, Ba, Be, Bi, Ca, Cd, Co, Cr, Cu, Fe, K, Li, Mg, Mn, Mo, Na, Ni, P, Pb, Rb, S, Sb, Se, Sr, Th, Ti, Tl, U, V and Zn	Wheat Triticum vulgare, sort Umanka—terrestrial	Concentrations of Ag, Cd, Cu, Pb, Sb, and Zn in the initial contaminated soil were 3–6 times higher than those in the initial clean soil. In particular, contents of Cu, Mo, Ni, Pb, Sb and Zn in roots of the wheat grown in the contaminated soil were higher than those in the roots of the plants grown in the clean soil. Moreover, all the elements except Pb transferred more easily from roots to leaves.
(28)	[66]	Field experiment (155 days (May–November))	Soil (agricultural soil)	Cd, Cr, Pb, As, and Hg	Rice (Oryza sativaL.)—terrestrial	The results showed the rice grain contained significantly lower amounts of five metals than straw and root in all sampling sites. Rice root accumulated Cd, As, and Hg from the paddy soil. The rice plant transported As very weakly, whereas Hg was transported most easily into the straw and grain among studied heavy metals.

No. (7), (10), (12), (18), (22), (23) adapted from no. (20). Phytoremediation Bibliography, Annotated Bibliography on Phytoremediation prepared by Mark Coleman, Biological Scientist, USDA Forest Service Southern Research Station and Ronald S.

Zalesny Jr., Research Plant Geneticist, USDA Forest Service North Central Research Station May 1, 2006.

Table 3: Phytoremediation study on water medium (hydroponic)

No.	Researcher	Research scale and duration	Uptake mechanisms and media (substrate)	Contaminant or parameter and concentration	Plants	Result
(1)	[67]	Field study (October–July 2005)	Water of Tasik Chini	Cd, Cu, and Pb	Five aquatic plant species, Lepironia articulata, Pandanus helicopus, Scirpus grossus, Cabomba furcata, andN-elumbo nucifera—aquatic	The highest concentration of heavy metals among the aquatic plants and plant parts was found in the roots of S. grossus. The concentrations of Cd in the leaves and stems of submerged aquatic plant, C. furcata, were higher than that in the leaves and stems of emergent aquatic plant and floating leaf plant. The concentration of Cu in the stem of C. furcata was greater than that in the leaf, while the concentration of Cd was more in the leaf than that in the stem. The heavy metal contents of the aquatic plants were in descending order of Pb > Cu > Cd. The highest internal translocation was found in P. helicopus, while the lowest internal translocation was found in S. grossus.

(2)	[68]	Laboratory (pot experiment)/14 days	Hydroponic	As and Se as $Na_2HAsO_4{\cdot}7H_2O$ and Na_2SeO_3/0, 0.73, 2.5, 4.27, 5.00 mg/L	Chinese brake fern (Pteris vittata L.)—terrestrial	At low levels of Se, As enhanced both Se uptake and the translocation of Se from roots to fronds. At higher levels of Se, As suppressed the uptake of Se. These results suggest that As serves to both stimulate and suppress Se uptake. The result is also in agreement with the well-known fact that Se is an element with both beneficial and toxic properties. The effect can change from beneficial to toxic based on the concentration of Se in plants.
(3)	[47]	Laboratory and Field study: wetland-pond system (Laboratory scale: 3 days cultivative and 84 hours exposure)	Field study: contaminated soil. Medium of laboratory scale experiment: L 0.1% Hoagland solution	Field study: Zn, Cu, Cd, and Pb. Laboratory: $ZnCl_2$, $CuCl_2$, $CdCl_2$, and $Pb(NO_3)_2$ (mixture of 20 μmol Zn, 0.5 μmol Cu, 1.5 μmol Cd, and 1.5 μmol Pb/L)	Potamogeton natans L.-aquatic Lemna gibba L.-aquatic Alisma plantago-aquatica L.-aquatic Sagittaria sagittifolia L.-aquatic Juncus effusus L-aquaticLemna minor L.-aquatic Elodea canadensis Michx.-aquatic Lythrum salicariaL.-aquaticPhalaris arundinacea L.-aquatic Impatiens parviflora DC.-terrestrial Urtica dioica L.—terrestrial Filipendula ulmaria L.-aquatic P. natans-aquatic A. plantago-aquatica-aquatic F. ulmarina-aquatic	The aquatic plants seem to have a higher metal accumulation capacity in shoots than terrestrial plants. This may be due to the capacity of aquatic plants to take up by shoot directly from the water. When submersed and free-floating plants are actively growing and accumulating metals directly from the water, they will function as an effective filter in stormwater treatment. Emergent plants in general mediate the binding of these metals in the sediment. Also, the terrestrial plants have the capacity to bind Cd and Zn to their roots, and; therefore, they can mediate a good stabilization of these metals in soil.

(4)	[32]	Laboratory (15 days)	Hydroponic	Pb as ($Pb(NO_3)_2$)	Indian mustard (Brassica juncea var. megarrhiza)—terrestrial	Brassica juncea is one plant which accumulates high levels of Pb and other heavy metals. The results indicate that lead nitrate obviously inhibits the root, hypocotyls, and shoot growth of Brassica juncea at the concentration of 10^{-3} M Pb^{2+}. Brassica juncea has the ability to accumulate Pb primarily in its roots, transport, and concentrate it in its hypocotyls and shoots in much lesser concentrations.
(5)	[30]	Laboratory (5 days)	Phytofiltration (water)	Mercury as $HgCl_2$ (0, 0.05, 0.5, 1, 2.5, 5, 10 mg/L)	Indian mustard (Brassica juncea)—terrestrial	Roots-concentrated Hg 100–270 times (on a dry weight basis) above initial solution concentrations. Mercury was more toxic to plants at 5 and 10 mg/L. The plants translocated little Hg to the shoots, which accounted for just 0.7–2% of the total Hg in the plants. Most Hg volatilisation occurred from the roots. Volatilised Hg was predominantly in the Hg(0) vapour form. Volatilisation was dependant on root uptake and absorption of Hg from the ambient solution. Efficiency process >95%.

(6)	[69]	Laboratory	Hydroponic	Arsenate (As(V)) and dimethylarsinic acid (DMAA)	Duckweed (Spirodela polyrhiza L.)—aquatic	The results show that not only internalized, but also surface-adsorbed arsenic (mostly arsenate) contributes significantly to the total amount of arsenic uptake in aquatic macrophyte S. polyrhiza L. The arsenic uptake in S. polyrhiza L. occurred through the phosphate uptake pathway as well as by physicochemical adsorption on Fe plaques of plant's surfaces. The arsenate uptake in the plant is related to the Fe ion and phosphate concentrations in culture medium while DMAA was not.
(7)	[2]	Laboratory (the contact times of 25–200 min were selected for the metal solutions (Co = 1.00 mM) with 2.0 g biomass/l at the obtained optimal pHs for each metal ion from the previous study)	Adsorption (water)	The Hg^{2+}, Cr^{3+}, $Cr^{6+}$and Cu^{2+} stock solutions were prepared by dissolving their corresponding salts, viz. $HgCl_2$, $CrCl_3{\cdot}3H_2O$, $K_2Cr_2O_7$, $CuCl_2$(analytical grade from Merck) in distilled water (pH values were almost 7.0, 5.0, 3.0 and 6.0 for Hg^{2+}, Cr^{3+}, Cr^{6+} and Cu^{2+}, respectively)	Lemna minor—aquatic	The potentiometric titration can be useful to study the pretreatment process of biomass (L. minor) using the acidic and alkali agents, the Qmax and KL values to remove Hg(II), Cr(III), Cr(VI), and Cu(II) from the aqueous solution by the activated L. minor at the alkali solution and by $CaCl_2/MgCl_2/NaCl$ with 1 : 1 : 1 molar ratio were higher than those for the reference one at the same conditions, the removal percents of metal ions by no. ACS L. minor was higher than ACS one at the pre-treatment pHs before 7.0, but it was higher by ACS biomass than no. ACS one at the pre-treatment pHs after 7.0.

(8)	[70]	Laboratory (seedling 2 weeks and treatment 2 weeks)	Hydroponic	Hg and Au (0, 50, 100, and 200 uMHg (as Hg $(CH_3COO)_2$) and 0 and 50 uMAu (as $KAuCl_4$) in hydroponics)	Chilopsis linearis(Cav.) sweet—terrestrial	The data showed that Au equimolar to Hg reduced the Hg toxicity. The concentration of Au and Hg in shoots indicated that C. linearis absorbed and translocated both Au and Hg at higher concentrations, compared to reported data. The data showed that the treatments produced structural alterations in both the vascular cylinder and the cortex. At the highest concentration, Hg produced a breakdown of the spongy parenchyma.
(9)	[71]	Laboratory (30 days)	Phytoextraction (water)	Mercury as $HgSO_4$ (0, 0.5 and 2 mg/L)	Water hyacinth (Eichornia crassipes)—aquatic Water lettuce (Pistia stratiotes)—aquatic Zebra rush (Scirpus tabernaemontani)—semi aquatic Taro (Colocasia esculenta)—aquatic	The higher the mercury concentration, the greater the amount of mercury removed by the plants. The largest uptake and accumulation capability is for water lettuce, followed by water hyacinth, taro and rush, respectively.
(10)	[72]	Laboratory—(pot experiment (10 days))	Hydroponic	As and Se (0, 150, or 300 uM of arsenat ($Na_2HAsO_4{\cdot}7H_2O$) in the presence of 0, 5 or 10 uM of selenat (Na_2SeO_4))	Pteris vitatta L.—terrestrial	Application of 5 uM Se enhanced As concentration by P. vittata fronds by 7–45%. At 5 uM, Se acted as an antioxidant, inhibiting lipid peroxidation (reduced by 26–42% in the fronds) via increased levels of thiols and glutathione (increased by 24% in the fronds). The results suggest that Se is either an antioxidant, or it activates plant protective mechanisms, thereby alleviating oxidative stress and improving arsenic uptake in P. vittata.

(11)	[45]	Laboratory (72 hours (for kinetics of Arsenic uptake), 3 days (effects of plant density, Plant reuse, and plant age), 10 days (groundwater remediation))	The groundwater was collected from a location which may have been contaminated from application of arsenical herbicides in the past.	As (pH 7.0, total As of 46 μg/L, As^{3+} of 1.6 μg/L, and total P of 20 μg/L)	Chinese Brake fern (Pteris vittata L.) plants—terrestrial	Chinese brake fern was efficient in taking up arsenic from a contaminated groundwater and was capable of reducing arsenic concentrations in the groundwater. One plant was sufficient to reduce arsenic in 600 mL groundwater to below 10 μg/L in 3 days. Young fern plants were more effective in arsenic removal than old fern plants of similar size. Ferns can be reused to remove arsenic from groundwater, but at a slower rate given the interval between exposures and nutritional status.
(12)	[73]	Laboratory	Hydroponic	Cu and Ni	Salix viminalis-clones and the basket willow Black Maul (S. triandra). S. burjatica "Germany", S.x dasyclados, S. candida and S. spaethii—terrestrial	The more resistant clones produced more biomass in the glasshouse and field and had higher metal concentrations in the wood. The less resistant clones had greater concentrations of Cu and Ni in the bark and produced less biomass in the glasshouse and field. Significant relationships were found between the response of the same clones grown in the short-term glasshouse hydroponics system and in the field.

(13)	[74]	Laboratory (10 days cultivate and 7 days exposure)	Nutrient solution	As (0, 5, 10, 20, 40 and 80 uM)	Azolla: A. caroliniana andA. filiculoides—aquatic	The efflux of arsenate was much higher (by about 9-fold) than that of arsenite. This may be because most of arsenite inside the cells was complexed with thiol compounds. The high As-accumulating Azolla (A.caroliniana) released approximately two times more As than the low-As accumulating Azolla (A. filiculoides). It appears that the amount of As efflux was proportional to the amount of As accumulation in the two strains of Azolla.

No. (12) adapted from no. (20). Phytoremediation Bibliography, Annotated Bibliography on Phytoremediation prepared by Mark Coleman, Biological Scientist, USDA Forest Service Southern Research Station and Ronald S. Zalesny Jr., Research Plant Geneticist, USDA Forest Service North Central Research Station, May 1, 2006.

Based on the collected data from the phytoremediation research listed in Tables 2 and 3, the accumulation of heavy metals As, Pb, and Hg in plant tissue is summarized in respective, Figures 4, 5, and 6.

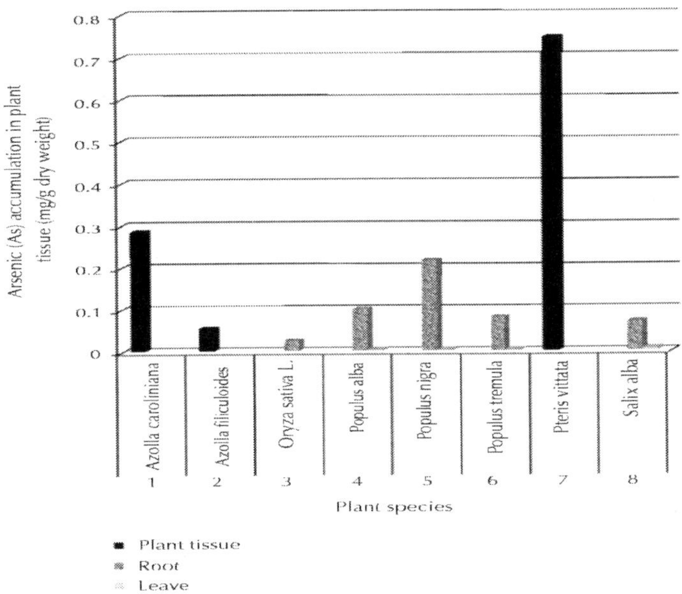

Figure 4: As accumulation in plant tissue.

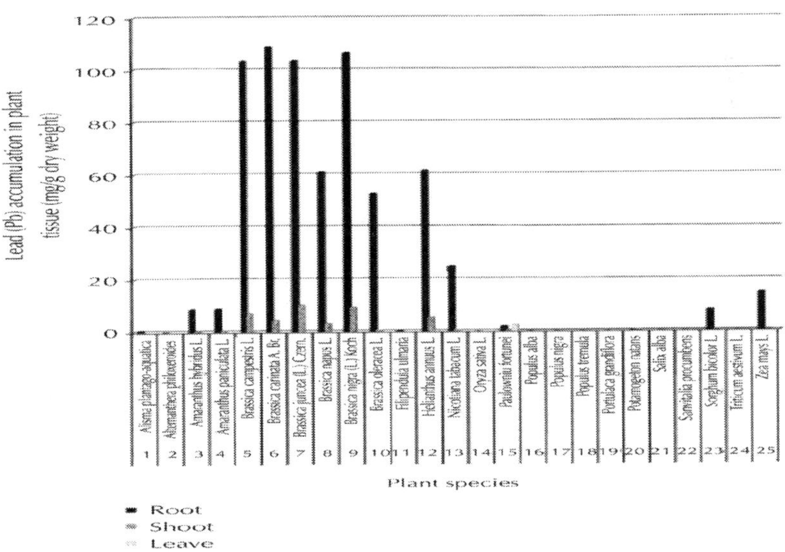

Figure 5: Pb accumulation in plant tissue.

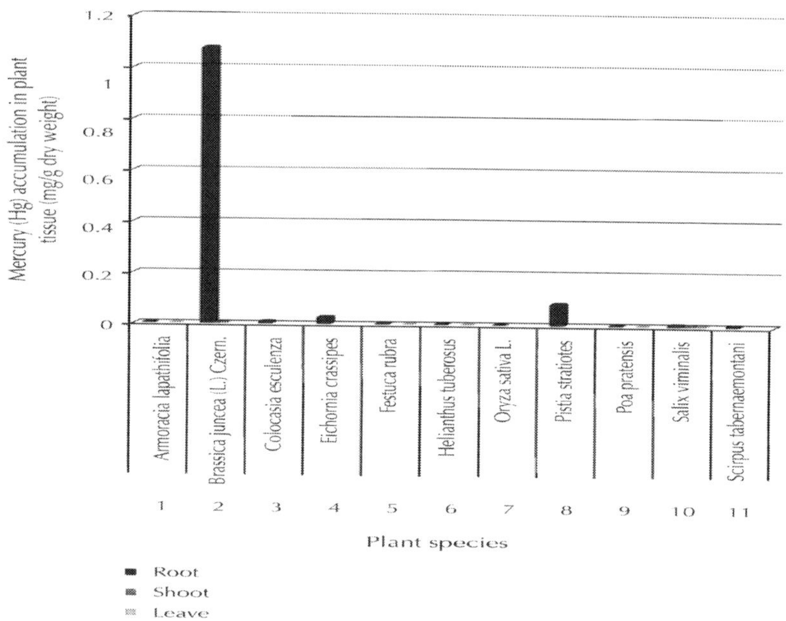

Figure 6: Hg accumulation in plant tissue.

According to Figure 4, the highest accumulation of As in plant tissue (the researchers have not detailed which part it is, but it might be the whole plant) occurs in Pteris vittata L. species. It can reach more than 0.7 mg As/g dry weight of plant. In plant root, the highest accumulation of As is in Populus nigra, which can reach more than 0.2 mg As/g dry weight of plant root.

As can be seen in Figure 5, several plants could accumulate Pb in their tissue of more than 50 mg/g dry weight of plant. Among those species are species of Brassica campestris L, Brassica carinata A. Br., Brassica juncea (L.)Czern. and Brassica nigra (L.) Koch that could accumulate more than 100 mg Pb/g dry weight.

Figure 6 shows that accumulated Hg in Brassica juncea L. Czern. is much higher than in other species of plants. It could reach more than 1 mg Hg/g dry weight of plant, while the other plants only accumulate less than 0.2 mg Hg/g dry weight.

ADVANTAGES OF PHYTOREMEDIATION

Phytoremediation techniques may also be more publicly acceptable, aesthetically pleasing, and less disruptive than the current techniques of physical and chemical process [38]. Advantages of this technology are its effectiveness in contaminant reduction, low-cost, being applicable for wide range of contaminants, and in overall it is an environmental friendly method. Figure 7 simplifies some advantages of phytoremediation technology.

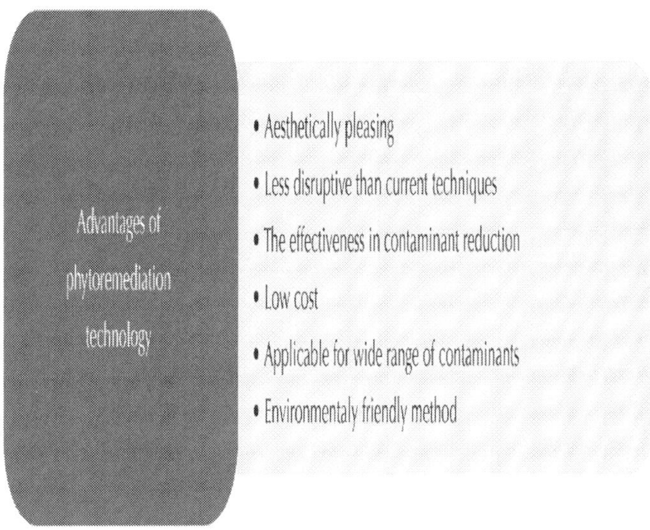

Figure 7: Advantages of phytoremediation technology.

The major advantages of the heavy metal adsorption technology by biomass are its effectiveness in reducing the concentration of heavy metal ions to very low levels and the use of inexpensive biosorbent materials [2]. Phytoremediation as possibly the cleanest and cheapest technology can be employed in the remediation of selected hazardous sites [29]. Phytoremediation encompasses a number of different methods that can lead to contaminant degradation [24].

Phytoremediation is a low-cost option and inexpensive approach for remediating environmental media, particularly suited to large sites that have relatively low levels of contamination [34]. This technology has been receiving attention lately as an innovative, cost-effective alternative to the more established treatment methods used at hazardous waste sites [29]. Phytoremediation potentially offers unique, low cost solutions to many currently problems of soil contamination [32, 75]. It is inexpensive (60–80% or even less costly) than conventional physicochemical methods, since it does not require expensive equipment or highly specialized personnel. It is cost-effective for large volumes of water having low concentrations of contaminants and for large areas having low to moderately contaminated surface soils [46].

It is applicable to a wide range of toxic metals and radionuclides [32] and also useful for treating a broad range of environmental contaminants, including organic and inorganic contaminants [46].

Phytoremediation is regarded as a new approach for the cleanup of contaminated soils, water, and ambient air [34]. Phytoremediation research can also contribute to the improvement of poor soils such as those with high aluminum or salt levels [75]. It is applicable to a range of toxic metals and radionuclides, minimal environmental disturbance, elimination of secondary air or water-borne wastes, and public acceptance [32]. Phytoextraction is considered as an environmentaly friendly method to remove metals from contaminated soils in situ. This method can be used in much larger-scale clean-up operations and has been applied for other heavy metals [76]. It is an esthetically pleasing, solar-energy-driven cleanup technology and there is minimal environmental disruption and in situ treatment preserves topsoil. In Situ applications decrease the amount of soil disturbance compared to conventional methods. It can be performed with minimal environmental disturbance with topsoil left in a usable condition and may be reclaimed for agricultural use. The organic pollutants may be degraded to CO_2 and H_2O, removing environmental toxicity [46]. Phytoremediation can be an alternative to the much harsher remediation technologies of incineration, thermal vaporization, solvent washing, or other soil washing techniques, which essentially destroy the biological component of the soil and can drastically alter its chemical and physical characteristics as well as creating a relatively nonviable solid waste. Phytoremediation actually benefits the soil, leaving an improved, functional soil ecosystem at costs estimated at approximately one-tenth of those currently adopted technologies [3]. It is the most ecological cleanup technology for contaminated soils and is also known as a green technology.

Another advantage of phytoremediation is the generation of a recyclable metal-rich plant residue [32]. Phytoremediation could be a viable option to decontaminate heavy-metal-polluted soils, particularly when the biomass produced during the phytoremediation process could be economically valorized in the form of bioenergy. The use of metal-accumulating bioenergy crops might be suitable for this purpose. If soils, contaminated with heavy metals, are phytoremediated with oil crops, biodiesel production from the resulting plant oil could be a viable option to generate bioenergy [34]. In large-scale applications, the potential energy stored can be utilized to generate thermal energy [46]. The success of the phytoextraction technique depends upon the identification of suitable plant species that can hyperaccumulate heavy metals and produce large amounts of biomass using established crop production and management practices [24].

LIMITATIONS OF PHYTOREMEDIATION TECHNOLOGY

On the other hand, there are certain limitations to phytoremediation system (Figure 8). Among them are being time-consuming method, the amount of produced biomass, the root depth, soil chemistry and the level of contamination, the age of plant, the contaminant concentration, the impacts of contaminated vegetation, and climatic condition.

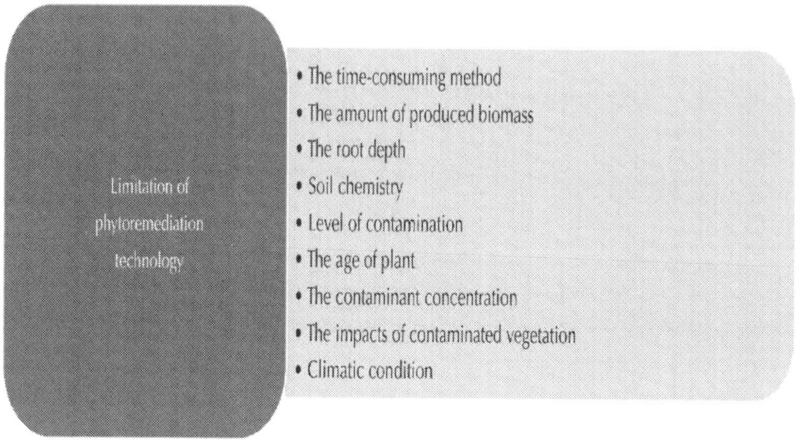

Figure 8: The limitation of phytoremediation technology.

Phytoremediation can be a time-consuming process, and it may take at least several growing seasons to clean up a site. The intermediates formed from those organic and inorganic contaminants may be cytotoxic to plants [46]. Phytoremediation is also limited by the growth rate of the plants. More time may be required to phytoremediate a site as compared with other more traditional cleanup technologies. Excavation and disposal or incineration takes weeks to months to accomplish, while phytoextraction or degradation may need several years. Therefore, for sites that pose acute risks for human and other ecological receptors, phytoremediation may not be the remediation technique of choice [29, 46]. Phytoremediation might be best suited for remote areas where human contact is limited or where soil contamination does not require an immediate response [38].

Under the best climatic conditions, with irrigation and fertilization, total biomass productivities can approach 100 t/ha/y. One of the unresolved issues is the tradeoff between toxic element accumulation and productivity. In practice, a maximum harvestable biomass yield of 10 to 20 t/ha/y would be likely, particularly for heavy metal accumulating plants. These values for productivity of biomass and heavy metal content would limit annual toxic element removal capacity between about 10 and 400 kg/ha/y, depending on the pollutant, plant species, climatic and other factors. For a target soil depth of 30 cm (4,000 t/ha), this amounts to an annual reduction from 2.5 to 100 ppm in soil toxic element levels. This is often an acceptable rate of contaminant removal, allowing site remediation over a few years to a couple of decades, particularly where the concentration of the contaminant can be lowered sufficiently to meet regulatory criteria. These values for productivity of biomass and heavy metal content would limit annual toxic element removal capacity between 10 and 400 kg/ha/y, depending on the pollutant, plant species, climatic and other factors [37].

The success of phytoremediation may be limited by factors such as growing time, climate, root depth, soil chemistry, and level of contamination [38]. Root contact is a primary limitation on phytoremediation applicability. Remediation with plants requires that contaminants be in contact with the root zone of the plants. Either the plants must be able to extend roots to the contaminants, or the contaminated media must be moved to be within range of the plants [29]. Restricted to sites with shallow contamination within rooting zone of remediative plants, ground surface at the site may have to be modified to prevent flooding or erosion [46].

Age greatly affects the physiological activity of a plant, especially its roots. Generally, roots of a young plant display greater ability to absorb ions than do those of an old plant when they are similar in size. It is important to use healthy young plants for more efficient plant removal. However, this does not rule out the use of larger older plants whose larger size may compensate for their lower physiological activity as compared to smaller younger plants [45].

High concentrations of contaminants may inhibit plant growth and, thus, may limit application on some sites or some parts of sites. This phytotoxicity could lead to a remedial approach in which high-concentration waste is handled with expensive ex situ techniques that quickly reduce acute risk, while in situ phytoremediation is used over a longer period of time to clean the high volumes of lower contaminant concentrations [29]. A major limitation in the phytoremediation of toxic elements is the maximal level that can be accumulated by plants. Plants with the highest levels of toxic metal contents, known as "hyperaccumulators", generally exhibit, on a dry weight basis, from about 2000 ppm (0.2%) for more toxic elements (Cd, Pb) to above 2% for the less toxic ones (Zn, Ni, Cu) [75]. Restricted to sites with low contaminant concentrations, the treatment is generally limited to soils at one meter from the surface and groundwater within a few meters of the surface with soil amendments may be required [46].

Some ecological exposure may occur whenever plants are used to interact with contaminants from the soil. The fate of the metals in the biomass is a concern. Although some forms of phytoremediation involve accumulation of metals and require handling of plant material embedded with metals, most plants do not accumulate significant levels of organic contaminants. While metal-accumulating plants will need to be harvested and either recycled or disposed of in compliance with applicable regulations, most phytoremediative plants do not require further treatment or disposal [29]. Harvested plant biomass from phytoextraction may be classified as a hazardous waste, hence, disposal should be proper. Consumption of contaminated plant biomass is a cause of concern; contaminants may still enter the food chain through animals/insects that eat plant material containing contaminants [46].

Climatic or hydrologic conditions may restrict the rate of growth of plants that can be utilized. Introduction of nonnative species may affect biodiversity [46].

CONCLUSIONS

Heavy metals uptake, by plants using phytoremediation technology, seems to be a prosperous way to remediate heavy-metals-contaminated environment. It has some advantages compared with other commonly used conventional technologies. Several factors must be considered in order to accomplish a high performance of remediation result. The most important factor is a suitable plant species which can be used to uptake the contaminant. Even the phytoremediation technique seems to be one of the best alternative, it also has some limitations. Prolong research needs to be conducted to minimize this limitation in order to apply this technique effectively.

REFERENCES

1. A. Gaur and A. Adholeya, "Prospects of arbuscular mycorrhizal fungi in phytoremediation of heavy metal contaminated soils," Current Science, vol. 86, no. 4, pp. 528–534, 2004.
2. R. Rakhshaee, M. Giahi, and A. Pourahmad, "Studying effect of cell wall's carboxyl-carboxylate ratio change of Lemna minor to remove heavy metals from aqueous solution," Journal of Hazardous Materials, vol. 163, no. 1, pp. 165–173, 2009. · ·
3. R. R. Hinchman, M. C. Negri, and E. G. Gatliff, "Phytoremediation: using green plants to clean up contaminated soil, groundwater, and wastewater," Argonne National Laboratory Hinchman, Applied Natural Sciences, Inc, 1995, http://www.treemediation.com/Technical/Phytoremediation_1998.pdf.
4. I. Shtangeeva, J. V.-P. Laiho, H. Kahelin, and G. R. Gobran, "Phytoremediation of metal-contaminated soils. Symposia Papers Presented Before the Division of Environmental Chemistry," American Chemical Society, Anaheim, Calif, USA, 2004,http://ersdprojects.science.doe.gov/workshop_pdfs/california_2004/p050.pdf.
5. K. Cho-Ruk, J. Kurukote, P. Supprung, and S. Vetayasuporn, "Perennial plants in the phytoremediation of lead-contaminated soils," Biotechnology, vol. 5, no. 1, pp. 1–4, 2006. ·
6. M. M. Lasat, "Phytoextraction of metals from contaminated soil: a review of plant/soil/metal interaction and assessment of

pertinent agronomic issues," Journal of Hazardous Substance Research, vol. 2, no. 5, pp. 1–25, 2000.

7. E. Pehlivan, A. M. Özkan, S. Dinç, and S. Parlayici, "Adsorption of Cu_{2+} and Pb_{2+} ion on dolomite powder," Journal of Hazardous Materials, vol. 167, no. 1–3, pp. 1044–1049, 2009.

8. S. Roy, S. Labelle, P. Mehta et al., "Phytoremediation of heavy metal and PAH-contaminated brownfield sites," Plant and Soil, vol. 272, no. 1-2, pp. 277–290, 2005.

9. D. Mohan and C. U. Pittman Jr., "Arsenic removal from water/wastewater using adsorbents—a critical review," Journal of Hazardous Materials, vol. 142, no. 1-2, pp. 1–53, 2007.

10. National Ground Water Association, Copyright 2001. Arsenic. What you need to knowhttp://www.ngwa.org/ASSETS/A0DD107452D74B33AE9D5114EE6647ED/Arsenic.pdf.

11. U.S. Department of Health and Human Services, Public Health Service Agency for Toxic Substances and Disease Registry. Division of Toxicology and Environmental Medicine. Arsenic. 2005,http://www.baltimorehealth.org/info/ATSDR%20fact%20sheet.pdf.

12. H. Hasegawa, M. A. Rahman, T. Matsuda, T. Kitahara, T. Maki, and K. Ueda, "Effect of eutrophication on the distribution of arsenic species in eutrophic and mesotrophic lakes," Science of the Total Environment, vol. 407, no. 4, pp. 1418–1425, 2009.

13. WHO Regional Office for Europe, Air Quality Guidelines, chapter 6.1, Arsenic, Copenhagen, Denmark, 2nd edition, 2000, http://www.euro.who.int/document/aiq/6_1_arsenic.pdf.

14. P. Chutia, S. Kato, T. Kojima, and S. Satokawa, "Arsenic adsorption from aqueous solution on synthetic zeolites," Journal of Hazardous Materials, vol. 162, no. 1, pp. 440–447, 2009.

15. H. A. Andrianisa, A. Ito, A. Sasaki, J. Aizawa, and T. Umita, "Biotransformation of arsenic species by activated sludge

and removal of bio-oxidised arsenate from wastewater by coagulation with ferric chloride," Water Research, vol. 42, no. 19, pp. 4809–4817, 2008.

16. R. J. Ampiah-Bonney, J. F. Tyson, and G. R. Lanza, "Phytoextraction of arsenic from soil by Leersia oryzoides," International Journal of Phytoremediation, vol. 9, no. 1, pp. 31–40, 2007.
17. M. Vaclavikova, G. P. Gallios, S. Hredzak, and S. Jakabsky, "Removal of arsenic from water streams: an overview of available techniques," Clean Technologies and Environmental Policy, vol. 10, no. 1, pp. 89–95, 2008.
18. A. M. Yusof and N. A. N. N. Malek, "Removal of Cr(VI) and As(V) from aqueous solutions by HDTMA-modified zeolite Y," Journal of Hazardous Materials, vol. 162, no. 2-3, pp. 1019–1024, 2009.
19. WHO Regional Office for Europe, Air Quality Guidelines, chapter 6.7, Lead, Copenhagen, Denmark, 2nd edition, 2001, http://www.euro.who.int/document/aiq/6_7lead.pdf.
20. J. H. Traunfeld and D. L. Clement, "Lead in Garden Soils. Home and Garden," Maryland Cooperative Extention, University of Maryland, 2001, http://www.hgic.umd.edu/_media/documents/hg18.pdf.
21. European Commission DG ENV. E3. Heavy Metals in Waste, Final Report Project ENV.E.3/ETU/2000/0058, 2002http://ec.europa.eu/environment/waste/studies/pdf/heavy_metalsreport.pdf.
22. J. F. Musselman and QEP, "Sources of Mercury in Wastewater, Pretreatment corner," http://www.cet-inc.com/cmsdocuments//7%20-%20Sources%20of%20Mercury%20in%20Wastewater%20(0204).pdf.
23. T. C. Chang, S. J. You, B. S. Yu, C. M. Chen, and Y. C. Chiu, "Treating high-mercury-containing lamps using full-scale thermal desorption technology," Journal of Hazardous Materials, vol. 162, no. 2-3, pp. 967–972, 2009.

24. L. Rodriguez, F. J. Lopez-Bellido, A. Carnicer, F. Recreo, A. Tallos, and J. M. Monteagudo, "Mercury recovery from soils by phytoremediation," in Book of Environmental Chemistry, pp. 197–204, Springer, Berlin, Germany, 2005.
25. I. Wagner-Döbler, "Microbiological treatment of mercury-loaded waste water—the biological mercury-decontamination-system. German Research Centre for Biotechnology,"http://www.gbf.de/mercury_remediation1/pdf-documents/Information%20leaflet.PDF.
26. A. Resaee, J. Derayat, S. B. Mortazavi, Y. Yamini, and M. T. Jafarzadeh, "Removal of Mercury from chlor-alkali industry wastewater using Acetobacter xylinum cellulose," American Journal of Environmental Sciences, vol. 1, no. 2, pp. 102–105, 2005.
27. A. Sas-Nowosielska, R. Galimska-Stypa, R. Kucharski, U. Zielonka, E. Małkowski, and L. Gray, "Remediation aspect of microbial changes of plant rhizosphere in mercury contaminated soil,"Environmental Monitoring and Assessment, vol. 137, no. 1–3, pp. 101–109, 2008. · ·
28. A. Erakhrumen and A. Agbontalor, "Review Phytoremediation: an environmentally sound technology for pollution prevention, control and remediation in developing countries," Educational Research and Review, vol. 2, no. 7, pp. 151–156, 2007.
29. U. S. Environmental Protection Agency, "Introduction to Phytoremediation," National Risk Management Research Laboratory, EPA/600/R-99/107, 2000, http://www.clu-in.org/download/remed/introphyto.pdf.
30. F. N. Moreno, C. W. N. Anderson, R. B. Stewart, and B. H. Robinson, "Phytofiltration of mercury-contaminated water: volatilisation and plant-accumulation aspects," Environmental and Experimental Botany, vol. 62, no. 1, pp. 78–85, 2008. ·
31. M. N. V. Prasad and H. M. De Oliveira Freitas, "Metal hyperaccumulation in plants—biodiversity prospecting forphytoremediation technology," Electronic Journal of Biotechnology, vol. 6, no. 3, pp. 110–146, 2003.

32. D. Liu, W. Jiang, C. Liu, C. Xin, and W. Hou, "Uptake and accumulation of lead by roots, hypocotyls and shoots of Indian mustard [Brassica juncea (L.)]," Bioresource Technology, vol. 71, no. 3, pp. 273–277, 2000. ·
33. T. Bhattacharya, D. K. Banerjee, and B. Gopal, "Heavy metal uptake by Scirpus littoralis Schrad. from fly ash dosed and metal spiked soils," Environmental Monitoring and Assessment, vol. 121, no. 1-3, pp. 363–380, 2006.
34. L. Van Ginneken, E. Meers, R. Guisson et al., "Phytoremediation for heavy metal-contaminated soils combined with bioenergy production," Journal of Environmental Engineering and Landscape Management, vol. 15, no. 4, pp. 227–236, 2007.
35. Interstate Technology and Regulatory Council, Phytotechnology Technical and Regulatory. Guidance and Decision Trees, 2009, http://www.itrcweb.org/guidancedocument.asp?TID=63.
36. R. K. Sinha, S. Herat, and P. K. Tandon, "14 phytoremediation: role of plants in contaminated site management," in Book of Environmental Bioremediation Technologies, pp. 315–330, Springer, Berlin, Germany, 2004.
37. U. S. Department of Energy, "Plume Focus Area, December. Mechanisms of plant uptake, translocation, and storage of toxic elements. Summary Report of a workshop on phytoremediation research needs," 1994,http://www.osti.gov/bridge/purl.cover.jsp;jsessionid=D72C8DD9003DCF51984EE254A6ED8BCB?purl=/10109412-BckU4U/webviewable/.
38. A. L. Salido, K. L. Hasty, J. M. Lim, and D. J. Butcher, "Phytoremediation of arsenic and lead in contaminated soil using Chinese Brake ferns (Pteris vittata) and Indian mustard (Brassica juncea),"International Journal of Phytoremediation, vol. 5, no. 2, pp. 89–103, 2003.
39. L. Erdei, G. Mezôsi, I. Mécs, I. Vass, F. Fôglein, and L. Bulik, "Phytoremediation as a program for decontamination of heavy-metal polluted environment," in Proceedings of the 8th Hungarian Congress on Plant Physiology and the 6th Hungarian Conference on Photosynthesis, 2005.

40. L. Erdei, G. Mezôsi, I. Mécs, I. Vass, F. Fôglein, and L. Bulik, "Phytoremediation as a program for decontamination of heavy-metal polluted environment," Acta Biologica Szegediensis, vol. 49, no. 1-2, pp. 75–76, 2005.
41. U. S. Environmental Protection Agency, Use of Field-Scale Phytotechnology, for Chlorinated Solvents, Metals, Explosives, and Propellants, and Pesticides Phytotechnology Mechanisms. Solid Waste and Emergency Response (5102G), EPA 542-R-05-002, 2005, http://www.clu-in.org/download/remed/542-r-05-002.pdf.
42. V. M. Ibeanusi, "Denise Antonia Grab In collaboration with Larry, Jensen Stephen Ostrodka—Environmental Protection Agency. Radionuclide Biological Remediation Resource Guide, U. S. Environmental Protection Agency," 2004, http://www.clu-in.org/download/remed/905b04001.pdf.
43. N. Merkl, R. Schultze-Kraft, and C. Infante, "Phytoremediation in the tropics—influence of heavy crude oil on root morphological characteristics of graminoids," Environmental Pollution, vol. 138, no. 1, pp. 86–91, 2005. · ·
44. J. G. Burken and J. L. Schnoor, "Phytoremediation: plant uptake of atrazine and role of root exudates,"Journal of Environmental Engineering, vol. 122, no. 11, pp. 958–963, 1996.
45. S. Tu, L. Q. Ma, A. O. Fayiga, and E. J. Zillioux, "Phytoremediation of arsenic-contaminated groundwater by the arsenic hyperaccumulating fern Pteris vittata L," International Journal of Phytoremediation, vol. 6, no. 1, pp. 35–47, 2004. · ·
46. W. J. S. Mwegoha, "The use of phytoremediation technology for abatement soil and groundwater pollution in Tanzania: opportunities and challenges," Journal of Sustainable Development in Africa, vol. 10, no. 1, pp. 140–156, 2008.
47. A. Fritioff and M. Greger, "Aquatic and Terrestrial Plant Species with Potential to Remove Heavy Metals from Stormwater," International Journal of Phytoremediation, vol. 5, no. 3, pp. 211–224, 2003.

48. P. Seuntjens, B. Nowack, and R. Schulin, "Root-zone modeling of heavy metal uptake and leaching in the presence of organic ligands," Plant and Soil, vol. 265, no. 1-2, pp. 61–73, 2004. ·
49. R. Chandra, R. N. Bharagava, S. Yadav, and D. Mohan, "Accumulation and distribution of toxic metals in wheat (Triticum aestivum L.) and Indian mustard (Brassica campestris L.) irrigated with distillery and tannery effluents," Journal of Hazardous Materials, vol. 162, no. 2-3, pp. 1514–1521, 2009.
50. M. Gupta, P. Sharma, N. B. Sarin, and A. K. Sinha, "Differential response of arsenic stress in two varieties of Brassica juncea L," Chemosphere, vol. 74, no. 9, pp. 1201–1208, 2009. · ·
51. D. Hammer, A. Kayser, and C. Keller, "Phytoextraction of Cd and Zn with Salix viminalis in field trials,"Soil Use and Management, vol. 19, no. 3, pp. 187–192, 2003.
52. R. E. Hamon, P. E. Holm, S. E. Lorenz, S. P. McGrath, and T. H. Christensen, "Metal uptake by plants from sludge-amended soils: caution is required in the plateau interpretation," Plant and Soil, vol. 216, no. 1-2, pp. 53–64, 1999.
53. M. S. Liphadzi, M. B. Kirkham, K. R. Mankin, and G. M. Paulsen, "EDTA-assisted heavy-metal uptake by poplar and sunflower grown at a long-term sewage-sludge farm," Plant and Soil, vol. 257, no. 1, pp. 171–182, 2003. ·
54. A. Murányi and L. Ködöböcz, "Heavy metal uptake by plants in different phytoremediation treatments," in Proceedings of the 7th Alps-Adria Scientific Workshop, Stara Lesna, Slovakia, 2008,http://www.mokkka.hu/publications/0387.117_MOKKA_AM_LK.pdf.
55. I. D. Pulford, D. Riddell-Black, and C. Stewart, "Heavy metal uptake by willow clones from sewage sludge-treated soil: the potential for phytoremediation," International Journal of Phytoremediation, vol. 4, no. 1, pp. 59–72, 2002.
56. J. Rydlová and M. Vosátka, "Effect of Glomus intraradices isolated from PB-contaminated soil on PB uptake by Agrostis

capillaris is changed by its cultivation in a metal-free substrate," Folia Geobotanica, vol. 38, no. 2, pp. 155–165, 2003.

57. L. Sebastiani, F. Scebba, and R. Tognetti, "Heavy metal accumulation and growth responses in poplar clones Eridano (Populus deltoides x maximowiczii) and I-214 (P. x euramericana) exposed to industrial waste," Environmental and Experimental Botany, vol. 52, no. 1, pp. 79–88, 2004. ·

58. S. Sharma, "Study on impact of heavy metal accumulation in Brachythecium populeum (Hedw.) B.S.G,"Ecological Indicators, vol. 9, no. 4, pp. 807–811, 2009. ·

59. T. Vamerali, M. Bandiera, L. Coletto, F. Zanetti, N. M. Dickinson, and G. Mosca, "Phytoremediation trials on metal- and arsenic-contaminated pyrite wastes (Torviscosa, Italy)," Environmental Pollution, vol. 157, no. 3, pp. 887–894, 2009. · ·

60. T. Vamerali, M. Bandiera, L. Coletto, F. Zanetti, N. M. Dickinson, and G. Mosca, "Phytoremediation trials on metal- and arsenic-contaminated pyrite wastes (Torviscosa, Italy)," Environmental Pollution, vol. 157, no. 3, pp. 887–894, 2009.

61. B. Vandecasteele, E. Meers, P. Vervaeke, B. D. Vos, P. Quataert, and F. M. G. Tack, "Growth and trace metal accumulation of two Salix clones on sediment-derived soils with increasing contamination levels,"Chemosphere, vol. 58, no. 8, pp. 995–1002, 2005.

62. P. Vervaeke, F. M. G. Tack, N. Lust, and M. Verloo, "Short- and longer-term effects of the willow root system on metal extractability in contaminated dredged sediment," Journal of Environmental Quality, vol. 33, no. 3, pp. 976–983, 2004.

63. M. Vyslouzilova, P. Tlustos, J. Szakova, and D. Pavlikova, "As, Cd, Pb and Zn uptake by Salix spp.clones grown in soil enrich by high load of this elements," Plant Soil Environment, vol. 49, no. 5, pp. 191–196, 2003.

64. H. B. Wang, Z. H. Ye, W. S. Shu, W. C. Li, M. H. Wong, and C. Y. Lan, "Arsenic uptake and accumulation in fern species growing at arsenic-contaminated sites of Southern China:

field surveys,"International Journal of Phytoremediation, vol. 8, no. 1, pp. 1–11, 2006.

65. J. Wang, C. B. Zhang, and Z. X. Jin, "The distribution and phytoavailability of heavy metal fractions in rhizosphere soils of Paulowniu fortunei (seem) Hems near a Pb/Zn smelter in Guangdong, PR China,"Geoderma, vol. 148, no. 3-4, pp. 299–306, 2009.

66. W.-X. Liu, L.-F. Shen, J.-W. Liu, Y.-W. Wang, and S.-R. Li, "Uptake of toxic heavy metals by rice (Oryza sativa L.) cultivated in the agricultural soil near Zhengzhou City, People›s Republic of China," Bulletin of Environmental Contamination and Toxicology, vol. 79, no. 2, pp. 209–213, 2007.

67. M. Ebrahimpour and I. Mushrifah, "Heavy metal concentrations (Cd, Cu and Pb) in five aquatic plant species in Tasik Chini, Malaysia," Environmental Geology, vol. 54, no. 4, pp. 689–698, 2008. ·

68. R. Feng, C. Wei, S. Tu, and X. Sun, "Interactive effects of selenium and arsenic on their uptake by Pteris vittata L. under hydroponic conditions," Environmental and Experimental Botany, vol. 65, no. 2-3, pp. 363–368, 2009. ·

69. M. A. Rahman, H. Hasegawa, K. Ueda, T. Maki, and M. M. Rahman, "Arsenic uptake by aquatic macrophyte Spirodela polyrhiza L.: interactions with phosphate and iron," Journal of Hazardous Materials, vol. 160, no. 2-3, pp. 356–361, 2008.

70. E. Rodríguez, J. R. Peralta-Videa, M. Israr et al., "Effect of mercury and gold on growth, nutrient uptake, and anatomical changes in Chilopsis linearis," Environmental and Experimental Botany, vol. 65, no. 2-3, pp. 253–262, 2009. ·

71. K. Skinner, N. Wright, and E. Porter-Goff, "Mercury uptake and accumulation by four species of aquatic plants," Environmental Pollution, vol. 145, no. 1, pp. 234–237, 2007.

72. M. Srivastava, L. Q. Ma, B. Rathinasabapathi, and P. Srivastava, "Effects of selenium on arsenic uptake in arsenic

hyperaccumulator Pteris vittata L," Bioresource Technology, vol. 100, no. 3, pp. 1115–1121, 2009.

73. C. Watson, I. D. Pulford, and D. Riddell-Black, "Screening of willow species for resistance to heavy metals: comparison of performance in a hydroponics system and field trials," International Journal of Phytoremediation, vol. 5, no. 4, pp. 351–365, 2003.

74. X. Zhang, A. J. Lin, F. J. Zhao, G. Z. Xu, G. L. Duan, and Y. G. Zhu, "Arsenic accumulation by the aquatic fern Azolla: comparison of arsenate uptake, speciation and efflux by A. caroliniana and A. filiculoides," Environmental Pollution, vol. 156, no. 3, pp. 1149–1155, 2008.

75. US Department of Energy, "Plume Focus Area. Summary Report of a Workshop on Phytoremediation Research Needs, Office of Technology Development, Office of Environmental Management and Division of Energy Biosciences, Office of Basic Energy Sciences, Office of Energy Research," 1994,http://www.law.csuohio.edu/lawlibrary-oldsite/info_services/acquisitions/acq-0104.html.

76. Y. Wang and M. Greger, "Use of iodide to enhance the phytoextraction of mercury-contaminated soil,"Science of the Total Environment, vol. 368, no. 1, pp. 30–39, 2006.

Chapter 4

Physical Property Relationships of the Rotokawa Andesite, a Significant Geothermal Reservoir Rock in the Taupo Volcanic Zone, New Zealand

Paul A Siratovich[1], Michael J Heap[2], Marlène C Villenueve[1], James W Cole[1], and Thierry Reuschlé[2]

[1]Department of Geological Sciences, University of Canterbury, Private Bag 4800, Christchurch 8140, New Zealand

[2]Laboratoire de Déformation des Roches, Équipe de Géophysique Expérimentale, Institut de Physique de Globe de Strasbourg (UMR 7516 CNRS, Université de Strasbourg/EOST), 5 rue René Descartes, Strasbourg cedex 67084, France

ABSTRACT

Background

Geothermal systems are commonly hosted in highly altered and fractured rock. As a result, the relationships between physical properties such as strength and permeability can be complex. Understanding such properties can assist in the optimal utilization of geothermal reservoirs. To resolve this issue, detailed laboratory studies on core samples from active geothermal reservoirs are required. This study details the results of the physical property investigations on Rotokawa Andesite which hosts a significant geothermal reservoir.

Methods

We have characterized the microstructure (microfracture density), porosity, density, permeability, elastic wave velocities, and strength of core from the high-enthalpy Rotokawa Andesite geothermal reservoir under controlled laboratory conditions. We have built empirical relationships from our observations and also used a classical micromechanical model for brittle failure. Further, we compare our results to a Kozeny-Carman permeability model to better constrain the fluid flow behavior of the rocks.

Results

We show that the strength, porosity, elastic moduli, and permeability are greatly influenced by pre-existing fracture occurrence within the andesite. Increasing porosity (or microfracture density) correlates well to a decreasing uniaxial compressive strength, increasing permeability, and a decreasing compressional wave velocity.

Conclusions

Our results indicate that properties readily measurable by borehole geophysical logging (such as porosity and acoustic velocities) can be used to constrain more complex and pertinent properties such as strength and permeability. The relationships that we have provided can then be applied to further understand processes in the Rotokawa reservoir and other reservoirs worldwide.

BACKGROUND

Fractures on multiple scales are the dominant control on fluid flow in most geothermal systems worldwide. Geothermal environments are prone to variable heat fluxes, dynamic fluid flow regimes, and active tectonics which impact the physical and mechanical properties of the reservoir rocks in which they are hosted. The influence of such a dynamic environment can render the host rocks highly altered, fractured, and microstructurally complex. As a result, the empirical correlation of physical properties to yield valuable relationships may not be entirely straightforward. Studies of these properties, and attempts to quantify how they relate to one another in the subsurface, can greatly assist in the optimization and maintenance of geothermal resources (e.g., Gupta and Sukanta [2006]; DiPippo [2008]; Grant and Bixley [2011]).

Here, we detail the results of a systematic physical and mechanical property study on the Rotokawa Andesite; the major reservoir unit within the high-enthalpy Rotokawa Geothermal Field (Krupp and Seward [1987]; Quinao et al. [2013]), located within the Taupo Volcanic Zone (TVZ), North Island, New Zealand (Figure 1). We first examine the texture, mineralogy, petrology, and microstructure. The key physical properties are then explicitly investigated: porosity, density, elastic wave propagation and dynamic elastic moduli, uniaxial compressive strength, static elastic moduli, and permeability. We empirically correlate the microcrack density of the andesite to the measured physical properties. Further, we present empirical relationships of physical properties and classical micromechanical and geometrical models to predict both uniaxial compressive strength and permeability, respectively. Our data is discussed in relation to the Rotokawa Geothermal Field and their applicability to other geothermal resources worldwide.

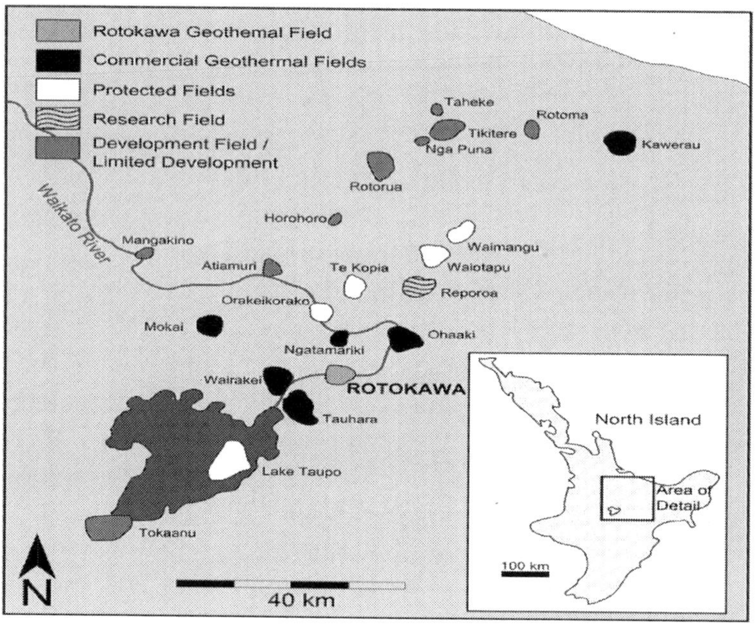

Figure 1: Geothermal fields of the Taupo Volcanic Zone (TVZ), North Island, New Zealand. Core used for this study was sourced from wells drilled in the Rotokawa Geothermal Field. (Adapted from Sewell et al. [2012]).

Previous Studies of Relevance

The study of the core from geothermal systems can yield valuable information to assist their modeling and understanding. For example, Stimac et al. ([2004]) present a study detailing the relationship between permeability and porosity from continuous core from Tiwi geothermal field, Philippines. Their data show that permeability and porosity decrease with depth, with occasional deviations attributed to alteration and compaction. However, the authors are careful to note that their work does not consider the influence of microfractures and their effect on relevant reservoir parameters. Lutz et al. ([2010]) present a case history of the well core from the Desert Peak field (NV, USA) in preparation for the stimulation of an enhanced geothermal system (EGS) by a thorough evaluation of petrological strength and elastic moduli. The results of their study elucidate relationships between clay mineralogy, rock fabric, and permeability increases as a result of mechanical shearing which support proposed hydraulic fracture operations in Well 27-15 at Desert Peak.

The effect of hydrothermal alteration on the physical properties of geothermal core is also a very significant area of research. Hydrothermal alteration can drastically change the elastic wave velocities and permeabilities of rock in both the natural and laboratory environment (Jaya et al.[2010]; Kristinsdóttir et al. [2010]; Pola et al. [2014]). However, coupled studies of physical properties such as porosity, permeability, and strength on geothermal reservoir rocks have not been extensively presented. A detailed study of the impact of a complex microstructure (microfractures and hydrothermal alteration) on the rock physical properties of a geothermal system such as Rotokawa could serve to greatly improve the understanding of reservoir processes at multiple scales.

Geothermal systems are more often than not associated with volcanic systems and are often hosted in rocks sourced from extinct volcanic systems. By proxy, the study of rocks from volcanic edifices can help to boost the understanding of processes within geothermal reservoirs especially with regard to microfractures, which play an essential role in controlling strength, porosity, permeability, elastic wave velocities, and elastic moduli of rocks (Wu et al. [2000]; Guéguen and Schubnel [2003]; Pereira and Arson [2013]; Faoro et al. [2013]; Pola et al. [2014]; Heap et al.[2014]). For example, Vinciguerra et al. ([2005]) studied the influence of thermal stressing on basaltic samples. They show, using elastic wave velocities, that the response of microstructurally variable basalts to thermal stressing can be quite different. While fresh microlitic basalt exhibited severe reductions in P-wave velocity after exposure to 900°C, the P-wave velocity of porphyritic basalt with a pervasive microcrack network did not change.

Similar dependence on the effect of microfractures on strength (Smith et al. [2009]) and permeability (Nara et al. [2011]) has been investigated, with microfractures proving to be deleterious to strength and to enhance permeability. Heap et al. ([2014]) showed, for a suite of pervasively fractured andesites, that an increase in porosity from 8 to 29 vol% decreases strength by a factor of 8 and increases permeability by 4 orders of magnitude. David et al. ([1999]) showed that mechanical and thermal microcracking in granites results in significant changes to permeability and elastic wave velocities. Mechanical microcracking resulted in the development of P-wave velocity anisotropy, while thermally microcracked samples showed little P-wave anisotropy. Additionally, permeability was much more varied in mechanically microcracked rocks than those induced thermally, suggesting that thermal microcracks develop isotropically. Chaki et al. ([2008]) investigated the role of thermal microcracking in granites and showed that elastic wave propagation is attenuated by microcracks and the orientation of these thermal microcracks (with regard to the original microstructure) plays a critical role in the propagation and attenuation of the waves. Faoro et al. ([2013]) provide a model for how microcrack density within an isotropically microcracked sample can be modeled as a function of aspect ratio and microcrack connectivity. Elastic moduli and elastic wave velocities are strongly influenced by the morphology, distribution, and shape of pore space in rocks and are substantially attenuated by the presence of microcracks (Stanchits et al. [2006] and references therein).

The relationship between porosity and strength has been observed by many authors, with general agreement that as the porosity of a sample (both rock and other engineering materials) increases, the strength decreases (e.g., Al-Harthi et al. [1999]; Li and Aubertin [2003]; Kahraman et al.[2005]; Chang et al. [2006]; Diamantis et al. [2009]; Ju et al. [2013]; Baud et al. [2014]; Heap et al. [2014]). The geometry of the pores also has a significant role in the strength of the materials both intrinsically and with respect to the direction of stress (Luping [1986]). The microstructure of rocks can be changed by increasing the crack damage (by mechanical and/or thermal stresses) as well as hydrothermal alteration (Heap et al. [2009]; Nara et al. [2011]; Pola et al. [2014]); these changes can be observed through the evaluation of destructive and nondestructive physical property measurements (Pola et al. [2012] and references therein; Sousa et al. [2005]). Further, Pola et al. ([2014]) also show that hydrothermal alteration of volcanic rocks can either strengthen or weaken rocks by decreasing or increasing their porosity, respectively.

Geological Significance of the Rotokawa Andesite

The TVZ is a rifted arc associated with the Hikurangi subduction system in which the Pacific plate descends beneath the Australasian plate (Cole [1990]; Wilson et al. [1995]), and hosts active volcanism and multiple associated hydrothermal systems (Bibby et al. [1995]; Rowland and Sibson[2004]; Rowland et al. [2010]). The Rotokawa field is one of these active hydrothermal systems and has been the subject of exploration for mineral resources (sulfur and gold deposits) and, for many years, was the subject of detailed investigation into its use as a commercial geothermal resource (Collar and Browne [1985]; Krupp and Seward [1987]; Hedenquist et al. [1988]). More recently, electricity generation has been realized at Rotokawa following the installations of the Rotokawa I (1997) and Nga Awa Purua (2010) generation stations (Legmann and Sullivan [2003]; Bloomberg et al. [2012]). The more recent of these installations, the Nga Awa Purua power station, hosts the single largest geothermal turbine installation in the world and has a generation capacity >140 MWe which is approximately 3 % of New Zealand's electricity consumption (Horie and Muto[2010]).

The main production zone for the installations at Rotokawa is from that of the Rotokawa Andesite, a series of lavas, pseudobreccias, and breccias. The movement of fluid through the andesite is predominantly along fracture networks (Rae [2007]; Massiot et al. [2012]). The andesite overlies basement of Miocene greywacke and is capped by a sequence of volcaniclastic and sedimentary units: Reporoa Group, Wairakei Ignimbrite, Waiora Formation, and Huka Falls Formation (Krupp and Seward [1987]; Rae [2007]). The andesite is gray to green and occasionally purple in color, depending on alteration within the reservoir; alteration is less intense in the lavas and more intense in the breccia and pseudo-breccia (Ramirez and Hitchcock [2010]). Production of reservoir fluids is sourced from the Rotokawa Andesite by 12 wells in the central part of the field (Figure 2), and re-injection of spent fluids is done through 5 wells along the southeastern margin of the field (Powell [2011]).

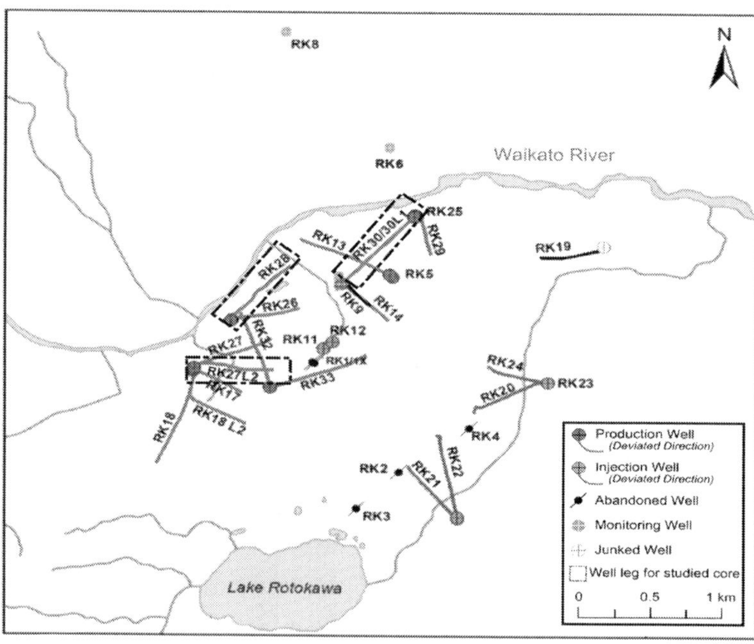

Figure 2: Rotokawa geothermal field and production and injection wells used within the field. Wells RK27L2, RK28, and RK30 were the source of the core used in this study and are outlined with dashed boxes in the figure.

METHODS

Study Source Material

The samples used in this study were sourced from Rotokawa production wells RK27L2, RK28, and RK30 (Figure 2). The measured depths (total borehole length measured from drilling rig floor), orientations, and corrected true vertical depths (TVD) are listed in Table 1. The original cores were approximately 6-m long and 100-mm diameter, and were initially described by the GNS Science Wairakei Research Centre, New Zealand, in a series of internal industry reports detailing the respective production wells from surface to total depth (TD). These reports describe the stratigraphic sequences of the wells and rock types, hydrothermal alteration, and locations of the wells (Rae et al. [2009]; Rae et al. [2010]; Ramirez and Hitchcock [2010]).

Table 1: Detail of core retrieval points from within the Rotokawa Andesite reservoir

Well name	Measured depth of core points (m)	True vertical depth (meters below reference level)	Inclination from vertical (degrees)	Azimuth from north (degrees)
RK28 ST1	2,310 to 2,316	−2,215 to 2,221	21.94	50.91
RK27 L2	2,120 to 2,126	−2,001 to 2,007	27.31	88.11
RK30 L1	2,320 to 2,326	−2,175 to 2,182	20.11	218.47

Measured depths are given as measured from the elevation of the drilling rig floor, true vertical depths are corrected to subsurface elevations, inclination is the deviation from vertical, and azimuth is the orientation of the borehole trajectory.

Siratovich et al.

Siratovich et al. Geothermal Energy 2014 2:10 doi:10.1186/s40517-014-0010-4

At the University of Canterbury (UC), the cores were catalogued and cut into workable cylinders approximately 100 mm in length. These smaller sections were over-cored to obtain smaller cylindrical samples 40 mm in diameter and ranging from 80 to 100 mm in length. All samples were machined so that their end faces were flat and parallel in accordance with ISRM standards (Ulusay and Hudson [2007]).

Microstructural characterization

The strength, porosity, permeability, and acoustic velocities can be significantly influenced by the presence of microfractures in a sample. Therefore, we deemed it necessary to develop a fundamental understanding of the microfracture densities in the samples. In order to characterize these features, 10 polished thin sections were prepared from offcuts of the cylinders used for the property characterization described below. The thin sections were prepared perpendicular to the core axis (X-Y plane) and, using reflected light thin section photomicrography (at ×40 magnification), were examined for microfracture densities using the methods suggested by Underwood ([1970]) and further described by Richter and Simmons ([1977]), Wu et al. ([2000]), and Heap et al. ([2014]). In each thin section, an 11 × 11 mm^2 area was selected, which was subdivided into sections of 1 × 1 mm^2. The number of cracks that intersected a grid array of parallel and perpendicular lines that were spaced at 0.1 mm was counted. This allowed the calculation of the crack surface area per unit volume according to Equation 1 (Underwood [1970]):

$$\mathrm{Sv} = \pi/2P_{\mathrm{l}} + (2 - \pi/2)P_{\mathrm{ll}} \quad (1)$$

where Sv is the crack surface area per unit volume (in mm^2/mm^3), P_{l} is the number of perpendicular lines crossed by crack intersections, and P_{ll} is the number of parallel lines crossed by crack intersections. We also characterized the anisotropy of microfracture distribution using Equation 2 (Underwood [1970]):

$$\Omega_{23} = P_{\mathrm{I}}-P_{\mathrm{II}}/P_{\mathrm{I}} + (4/\pi - 1)P_{\mathrm{II}} \tag{2}$$

Density and Porosity Measurements

Once the samples were cut and ground flat and parallel, they were washed with water to remove any debris from sample preparation. They were then immersed in distilled water under vacuum of about 100 kPa for 24 h. Samples were taken out of the water and were weighed after their surface water had been removed. The samples were then placed into a laboratory oven at 105°C and dried until a constant mass was observed. Subsequently, they were removed from the oven and held in a dessicator until further characterization was implemented. Sample lengths and diameters were measured to within 0.01 mm. The connected porosity and dry bulk density of the samples were calculated following the methods recommended by Ulusay and Hudson ([2007]).

Characterization of Elastic Wave Velocities and Dynamic Elastic Moduli

The compressional wave (Vp) and shear wave (Vs) velocities and dynamic elastic moduli were measured using a GCTS (Geotechnical Consulting and Testing Systems, Tempe, AZ, USA) Computer Aided Ultrasonic Velocity Testing System (CATS ULT-100) apparatus with axial P- and S-wave piezoelectric transducers (Figure 3). The resonance frequency of the transducers was 900 kHz, pulse acquisition rate was 20 MHz, and 108 waveforms were captured for each sample. The velocities were collected under a constant uniaxial stress of 10 MPa via a Tecnotest servo-controlled 3,000 kN loading frame (Technotest, Modena, Italy) (Figure 3). The stress of 10 MPa was used to ensure a consistent waveform across the specimens and that applied stress was consistent for all measurement cycles. This was determined to be below microcrack closure and opening stress by analyzing the change in axial strain as the sample was loaded to 10 MPa (Eberhardt et al. [1998]). There was no change in axial strain and absence of acoustic emissions (AEs) during the initial loading (Brace et al. [1966]; Martin and Chandler [1994]; Lion et al. [2005]; Nicksiar and Martin [2012]); this ensured a good quality interpretation of the first arrival time of elastic wave pulses. Using these data, we determined the dynamic Poisson's ratio and Young's modulus using Equations 3 and 4 (Guéguen and Palciauskas [1994]), respectively:

$$V_d = (Vp^2 - 2Vs^2)/2(Vp^2 - Vs^2) \tag{3}$$

$$E_d = (\rho Vs^2(3\,Vp^2 - 4Vs^2))/(Vp^2 - Vs^2) \tag{4}$$

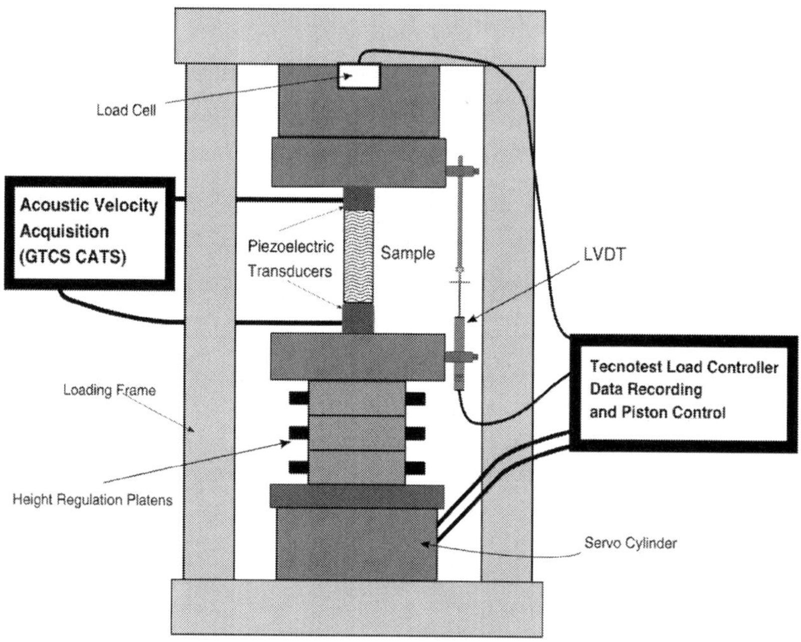

Figure 3: Loading frame set-up for acquisition of elastic velocities at the University of Canterbury. The frame is a Tecnotest 3,000 kN loading frame and a stress of 10 MPa was applied to each sample cycle to ensure a consistent waveform and quality picking of the first arrival time of the acoustic pulses (note that the figure is not to scale).

Where Vp is compressional wave velocity in meters per second, Vs is shear wave velocity in meters per second, E_d is the dynamic Young's modulus in pascal, V_d is the dynamic Poisson's ratio, and ρ is density in kilograms per cubic meter. Our physical property characterizations are summarized in Table 2. In addition to the determination of the elastic moduli from our elastic wave velocity measurements, we also utilized the method of Martínez-Martínez et al. ([2011]) to characterize the spatial attenuation of the compressional waveform anisotropy (Equation 5):

$$\alpha_s = 20 \log (A_e / A_{max}) / L \tag{5}$$

where α_s is spatial attenuation in decibels per centimeter, A_e is the maximum amplitude emitted by the piezoelectric crystal, A_{max} is the recorded maximum amplitude of the pulse after passing through the sample, and L is the length of the sample in meters.

Table 2: Results of quantitative microstructural characterization

Sample name	**Crack density for intercepts parallel to orientation axis$P$|| (mm−1)**	**Crack density for intercepts perpendicular to orientation axis$P$ | (mm−1)**	**Crack area per unit volume Sv (mm2/mm3)**	**Anisotropy factorΩ2,3**	**Connected porosity (vol%)**
27_21_0B	13.73	13.53	13.06	0.01	14.91
28_10_5A	4.10	4.41	8.33	0.06	7.47
27_20_4_B	1.83	2.06	3.77	0.08	4.37
27_3_3B	4.77	4.77	9.55	0.01	9.81
30_22_4A	2.77	2.62	5.48	0.01	6.49
28_10_9B	3.84	3.90	7.71	0.03	7.42
28_12_1	4.83	4.65	9.59	0.02	7.89
30_21_1B	2.93	2.91	5.85	0.01	7.47
27_21_3A	5.80	5.13	11.31	0.09	16.3
28_10_6C	2.97	2.87	6.38	0.05	5.97

As discussed in the section on microstructural characterization, crack densities were calculated on thin section samples to ascertain crack areas per unit volume using the optical microscope method (Underwood[1970]).

Siratovich et al.

Siratovich et al. Geothermal Energy 2014 2:10 doi:10.1186/s40517-014-0010-4

Uniaxial Compressive Strength Testing and Static Elastic Moduli

Uniaxial compressive strength (UCS) was determined using a Technotest 3,000 kN, servo-controlled loading frame (Figure 4). Four strain gauges (20-mm strain gauges with a gauge factor of 2.12 supplied by Tokyo Sokki Kenkyujo Co. Ltd. (TML) Shinagawa-ku, Tokyo, Japan) were glued onto each sample. Two vertical gauges measured axial strain and two laterally oriented gauges measured radial strain; care was taken to ensure that the strain gauges were perpendicular to their respective axes of deformation. The specimens were deformed at a constant strain rate of $1.0 \times 10^{-5}\ s^{-1}$ (controlled by linear variable differential transformer, LVDT) at ambient laboratory temperature and humidity conditions. During experimentation, AE output was monitored using Physical Acoustics Corporation MISTRAS' AE node acquisition system (Princeton Jct, NJ, USA). Two physical acoustics WS AE transducers (100 to 900 kHz operating frequency) were attached to the samples at the top and base, and hit counts, waveforms, energy, and amplitude of the received signals were recorded during sample deformation. AE monitoring was used during deformation as a proxy for microcracking as AEs are generated by the release of energy from a material during the propagation and nucleation of microcracks (Eberhardt et al. [1998]; [Diederichs et al. 2004]). We utilized arbitrary AE energy units (the area under the received waveform signal) for comparison of AE activity across the datasets. Once stress-strain curves were obtained and AE data are processed, we calculated the static elastic moduli for each specimen utilizing Equations 6 and 7 with the tangent deformation modulus at 50 % of the maximum peak stress (Ulusay and Hudson [2007]). In addition, we selected portions of the stress-strain sequence to identify crack closure, crack initiation, unstable crack propagation, and, ultimately, crack coalescence and sample failure (Martin [1993]; Eberhardt et al. [1998]; Takarli et al. [2008]; Heap and Faulkner[2008]):

$$E_s = (\Delta\sigma a/\Delta\varepsilon a) \tag{6}$$

$$\nu_s = -(\Delta\varepsilon r/\Delta\varepsilon a) \tag{7}$$

where E_s is the static Young's modulus (Pa), ν_s is the static Poisson's Ratio, σa is the differential axial stress (Pa), εa is the axial strain, and εr is the radial strain.

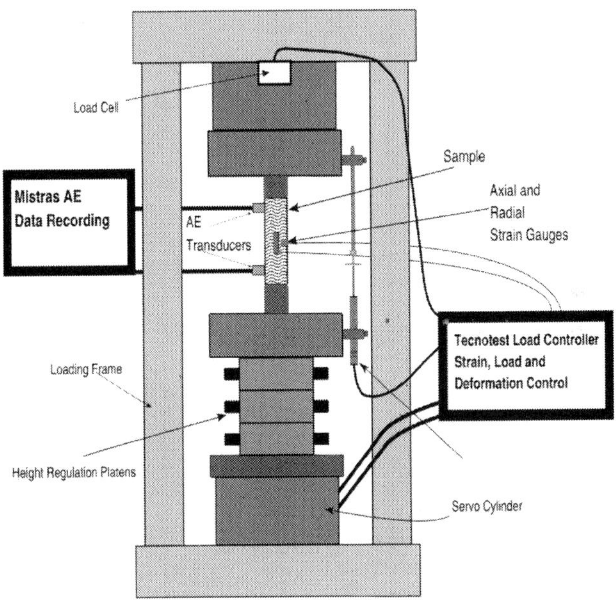

Figure 4: Loading frame set-up for determining uniaxial compressive strength (UCS) at University of Canterbury. Testing was carried out using TML strain gauges and MISTRAS acoustic emission

monitoring equipment. Loading during axial differential stressing was achieved through a constant strain rate of 1.0×10^{-5} s^{-1} (not to scale).

Permeability Measurements

Gas (argon) permeability measurements were made at the Laboratoire de Déformation des Roches, Université de Strasbourg (France). The 40-mm diameter specimens were over-cored to a diameter of 20 mm and cut and ground flat and parallel to a nominal length of 40 mm. The new samples were then re-evaluated by the triple-weight method to obtain porosity via the Archimedes' method (Ulusay and Hudson [2007]) and oven-dried under vacuum at 40°C until no change in sample mass was observed. The samples were then jacketed with viton sleeves, placed between two steel end-caps and lowered into the pressure vessel (Figure 5). A confining pressure of 2 MPa was applied to the sample (provided by distilled water), and permeability measured using the transient method (or pulse-decay method). For the permeability measurements, an initial differential pore pressure was applied to the sample, the upstream inlet was then closed, and the pore pressure decay monitored over time. The downstream fluid pressure (P_{down}) was the ambient atmospheric pressure, and the maximum upstream fluid pressure (P_{up}) was set so the pressure differential was 0.5 MPa. Permeability was then calculated using Equation 8 (after Brace et al.[1968]):

$$k_{gas} = (2\eta L/A)\left(V_{up}/(P_{up})^2 - (P_{down})^2\right)(\Delta P_{up}/\Delta t) \tag{8}$$

where k_{gas} is the gas permeability, is the viscosity of the pore fluid, A is the cross-sectional area of the sample, V_{up} is the volume of the upstream pore pressure circuit (approximately 7 cm^3), P_{up} is the upstream pore pressure, P_{down} is the downstream pore pressure, and t is the time. By plotting P_{up} as a function of time, the local slope of the curve is computed to determine the temporal variation of the permeability k_{gas}. To check whether our data should be corrected for Klinkenberg's 'slip flow' (Klinkenberg [1941]), we plotted the measured gas permeability as a function of the inverse of the mean pore fluid pressure, P_{mean}. For the transient method, since P_{down} is constant, the decay of P_{up} through time corresponds to the decay of the mean pore pressure P_{mean}. We found that, in all cases, the Klinkenberg correction should be applied:

$$k_{\mathbf{true}} = k_{\mathbf{gas}}(1 + b/P_{\mathbf{mean}}) \tag{9}$$

where k_{true} is the true gas permeability, b is Klinkenberg slip factor, and P_{mean} is the mean pore fluid pressure.

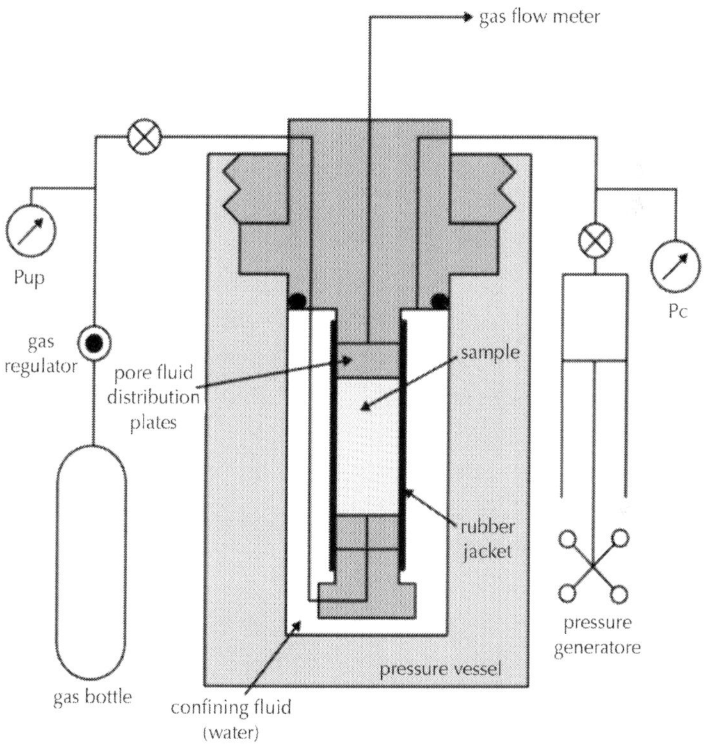

Figure 5: Gas permeameter used to measure permeability at University of Strasbourg (not to scale). Confining pressure of 2 MPa was applied using distilled water as the pressurizing media, and pore fluid was dry argon gas supplied at 1.5 MPa for a net effective pressure of 0.5 MPa.

RESULTS

In the following section, we present our data and observations on petrology, microstructure (quantitative microfracture analysis), macrostructure (bulk density, porosity, acoustic wave velocities, and dynamic moduli), strength relations (by UCS testing), and finally the ability of the rock to transmit fluid (permeability) of the Rotokawa Andesite.

Petrology

The Rotokawa Andesite shows moderate to intense hydrothermal alteration with the groundmass and phenocrysts showing replacement of original mineralogy. Fractures and occasional veins of quartz, calcite, anhydrite, and epidote occur, and amygdales within the sample are often filled with chlorite, calcite, hematite, pyrite, and chalcedony, often with quartz rims (Figure 6). Alteration is pervasive with the original mineral assemblages typically replaced by secondary hydrothermal alteration species, with some specimens showing very little original mineralogical texture. Plagioclase feldspars have been altered to albite, adularia, occasional calcite, and rare pyrite, and ferromagnesian minerals have been replaced by chlorite, quartz, calcite, and occasional epidote. Microfractured phenocrysts (Figures 6 and 7) are abundant and many relict phenocrysts retain original texture are but replaced by secondary mineralization. The average phenocryst size is 0.5 to 1 mm with occasional plagioclase near 1.5 to 2 mm; amygdales also range from 1 to 1.5 mm in size. The alteration chemistry of the samples indicates that this portion of the reservoir is dominated by chlorite/epidote alteration. The degree of alteration is relatively consistent across the core we have sampled with most primary mineralogies replaced by secondary alteration products. Microfracture mineralization indicates that these networks may have been conductive pathways for fluid migration (i.e., the presence of chlorite clays, adularization of plagioclase, calcite, and quartz rimming of fractured matrix); we typically observe chlorite, calcite, and quartz as alteration mineralogies with occasional epidote centers within the fractures. Backscatter scanning electron microscopy (SEM) was utilized to further reiterate the complex interaction of fractures and vesicles in the specimens (Figure 7). At several different magnifications, we see an abundance of microfractures in the samples as well as a clear depiction of the complex alteration mineralogy displayed by the andesite.

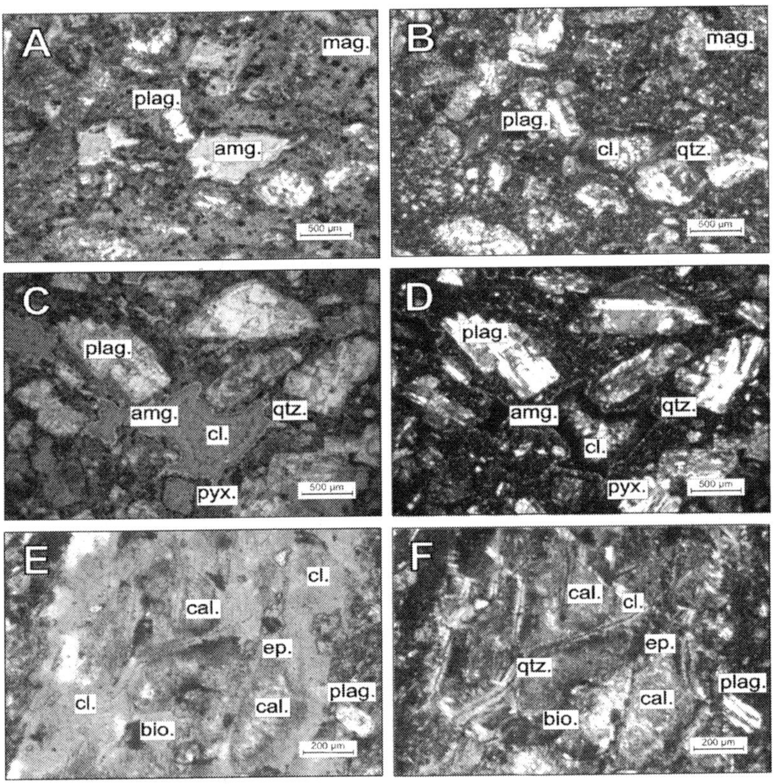

Figure 6: Thin section photomicrographs of the Rotokawa Andesite. (A) Plane-polarized light of RK28 2310.6C andesitic pseudo-breccia with plagioclase laths (plag.), groundmass is altered plagioclase, abundant magnetite (mag.), and amygdale (amg.). (B) Cross-polarized light of RK28 2310.6C clearly shows alteration fabrics of the brecciated andesite with plagioclase (plag.) and amygdales filled with chlorite (cl.) and rimmed by quartz (qtz.). (C) Plane-polarized light view of RK27_L2 2121.4A showing

andesitic breccia with plagioclase with slight adularia alteration (plag.) and amygdale (amg.) filled with chlorite (cl.), quartz (qtz.), and highly altered pyroxene (pyx.) in lower portion of image. (D) Cross-polarized light view of RK27_L2 2121.4A shows chlorite infill of a large amygdale (amg.) in the center of the photomicrograph and quartz rim (qtz.), plagioclase (plag.), and highly altered pyroxene (pyx.). (E) Plane-polarized light view of RK30 2322.4A shows highly altered and microfractured plagioclase phenocryst with intense alteration and replacement by chlorite (cl.), epidote (ep.), calcite (cal), possible biotite (bio.), and small plagioclase showing evidence of adularia alteration (plag.). (F) Cross-polarized light of RK30 2322.4A illustrates microfracture network and veining with alteration products of quartz (qtz.), epidote (ep.), biotite (bio.), calcite (cal.), and plagioclase (plag.).

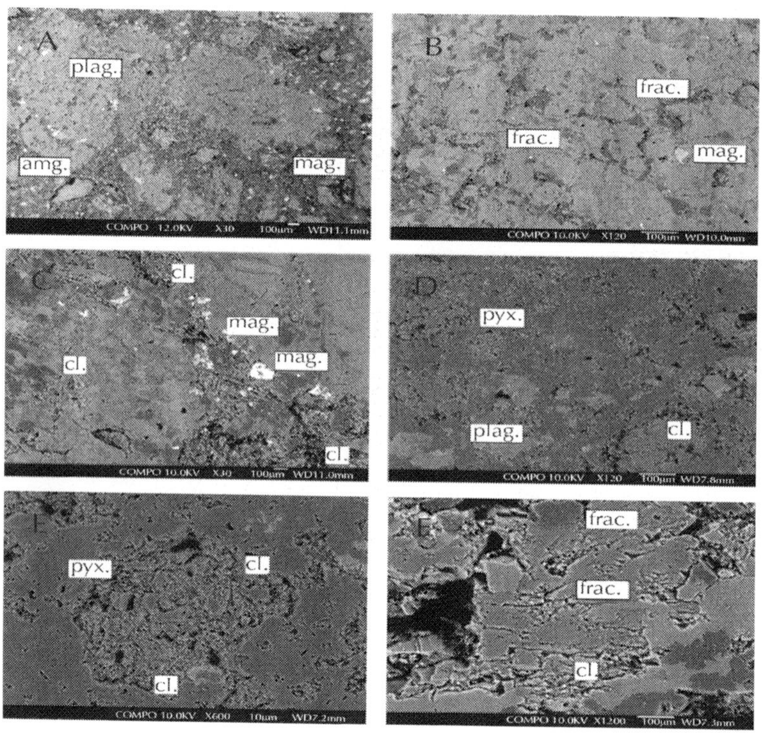

Figure 7: Backscattered scanning electron microscope photomicrographs of Rotokawa Andesite. Samples A to C are from RK28 at 2,310.6 m depth and D to F are from RK27 at 2,121.1 m depth.(A) Andesite with abundant

ferromagnesian minerals (mag.), altered amygdale (amg.), and highly scattered magnetites (mag. bright hues).(B) Detail of a fractured plagioclase phenocryst showing the microfractured texture (frac.) and occasional magnetite (mag.). (C)Detail of a fracture infill showing chloritization (cl.) and abundant ferromagnesian minerals (mag.). (D) Groundmass of RK27 sample; pervasive fracturing is not apparent at this magnification but the porous network is quite apparent with pyroxene (pyx.), chlorite (cl.), and plagioclase (plag.). (E) Replacement mineralogy of likely pyroxene phenocryst (pyx.) showing abundant chloritization (cl.) and dissolution textures. (F) Detail of relict pyroxene and abundant chlorite (cl.) with abundant microfractures (cl.) apparent in the sample mass.

Quantitative Two-Dimensional Microstructural Analysis

We evaluated the microfracture density of 10 specimens as a function of crack surface area per unit volume (Table 2). These samples were selected to represent the range of connected porosities observed within the sample set. We found that the crack area per unit volume in our samples ranges from 3.77 to 13.06 mm^2/mm^3 and appears to be independent of the alteration and mineralogy of the specimens. The calculated anisotropy factor ($\Omega_{2,3}$), indicates that the microcracks are isotropic (Table 2).

Porosity and Bulk Density

Bulk density decreases as connected porosity increases, as expected for samples of similar composition (Figure 8A). Bulk dry densities of the samples range from 2.29 to 2.65 g/cm^3, with a mean value of 2.49 g/cm^3. The connected porosities range from 4.37 to 16.3 vol%, with a mean value of 8.44 vol%. While there is some variation in the distribution of pores/vesicles in the samples, we observe that the microcrack density exerts an important control on the porosity and density, as illustrated by the correlation between crack area per unit volume and connected porosity presented as Figure 8B.

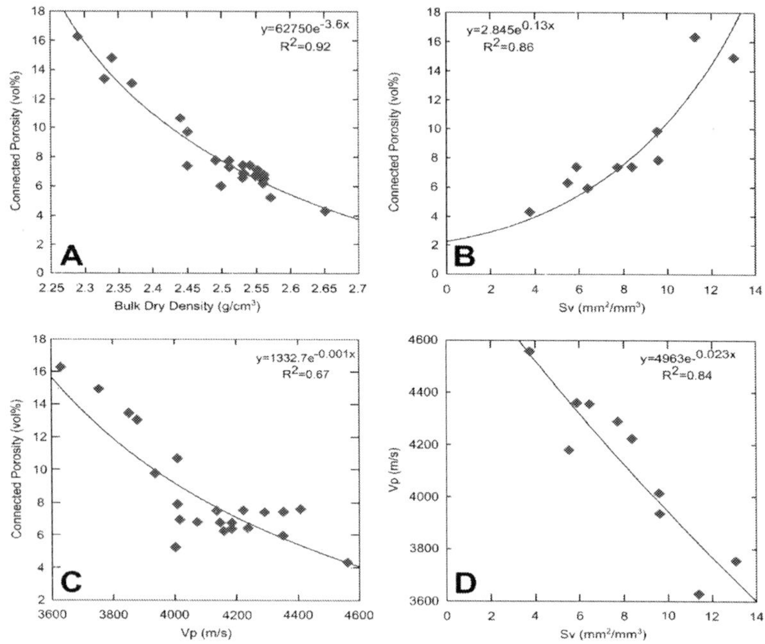

Figure 8: Relationships of porosity, density, crack area, and compressional wave velocity for Rotokawa Andesite. (A) Connected porosity versus dry bulk density for Rotokawa Andesite calculated using the dual weight method of (Ulusay and Hudson [2007]). (B) Crack area per unit volume (Sv) plotted versus connected porosity for samples representing the range of values measured within the measured dataset.(C) Connected porosity (vol%) using the dual weight method versus axial compressional wave velocity (Vp) as measured under a stress of 10 MPa which was determined to be below the crack closure stress (see text for further detail). (D) Crack area per unit volume (Sv) versus compressional wave velocity (Vp).

Ultrasonic Wave Velocities, Dynamic Elastic Moduli, and Spatial Attenuation

Measurements made on dry samples under ambient (pressure and temperature) conditions yielded axial P-wave velocities from 3,627 to 4,556 m/s with a mean value of 4,106 m/s, and axial S-wave velocities between 2,160 to 2,752 m/s with a mean value of

2,510 m/s (Table 3). Porosity and P-wave velocities show moderate correlation: samples with higher porosities have slower elastic wave velocities, as seen in Figure 8C. The crack area per unit volume also correlates well with P-wave velocity (Figure 8D). The axial spatial attenuation for the andesites ranges from 8.39 to 28.74 dB/cm (Figure 9). Figure 9 shows that there is no clear trend between spatial attenuation and P-wave velocity. Dynamic Poisson's ratio and dynamic Young's modulus were in the range of 0.13 to 0.23 and 24.6 to 45.9 GPa, respectively (Table 3).

Table 3: Physical property measurements of 22 samples used in destructive testing of Rotokawa Andesite

Sample source_ well sample name	Bulk dry density (g/cm3)	Connected porosity (vol%)	Vp (m/s)	Vs (m/s)	Spatial at- tenuation (dB/cm)	UCS (MPa)	Static Young's modulus (GPa)	Dynamic Young's modulus (GPa)	Static Poisson's ratio	Dynamic Poisson's ratio
RK_27_ L2_21.5B	2.44	10.72	4,005	2,443	14.63	85.99	19.9	35.1	0.24	0.2
RK_27_ L2_21.8A	2.33	13.49	3,850	2,363	16.47	79.91	25.2	31.2	0.26	0.2
RK_27_ L2_23.2A	2.56	6.61	4,182	2,490	22.48	105.26	31.2	38.9	0.19	0.23
RK_27_ L2_20.4B	2.65	4.37	4,556	2,752	22.53	211.05	37.7	45.9	0.25	0.21
RK_27_ L2_21.1C	2.37	13.1	3,877	2,405	23.6	69.53	21.5	32.5	0.18	0.19
RK_27_L2_3.3B	2.45	9.81	3,937	2,331	23.51	95.78	32.4	29.9	0.13	0.18
RK_27_ L2_21.0B	2.34	14.91	3,752	2,337	8.4	60.13	28.1	30.6	0.12	0.17
RK_27_ L2_21.3A	2.29	16.3	3,627	2,160	15.13	70.57	30.4	24.6	0.16	0.17
RK_28_10.6C	2.5	5.97	4,350	2,652	18.27	146.2	43.7	42.4	0.27	0.2
RK_28_10.8C	2.53	6.72	4,147	2,537	11.67	109.91	27.2	39.1	0.34	0.2
RK_28_10.9B	2.51	7.42	4,285	2,615	14.67	137.31	32.4	41.5	0.2	0.2

RK_28_13.2A	2.55	6.97	4,013	2,531	18.53	146.21	38.3	37.2	0.24	0.19
RK_28_10.5A	2.45	7.47	4,220	2,578	14.83	130.71	27.4	38.8	0.25	0.21
RK_28_11.5A	2.51	7.62	4,403	2,555	10.06	152.76	35.6	37.4	0.27	0.13
RK_28_12.1	2.49	7.89	4,010	2,460	12.23	115.01	29.3	36.2	0.22	0.14
RK_30_20.4A	2.57	5.3	4,002	2,495	10.39	140.97	33.6	36.9	0.09	0.15
RK_30_21.0A	2.55	6.84	4,070	2,508	11.17	126.53	26.4	39.8	0.14	0.15
RK_30_21.1B	2.54	7.47	4,352	2,659	28.75	157.93	25.8	43.2	0.17	0.2
RK_30_21.7B	2.56	6.28	4,154	2,588	19.91	162.71	28.3	41.1	0.22	0.18
RK_30_22.3B	2.53	7.51	4,133	2,550	21.72	137.97	31.5	39.2	0.23	0.19
RK_30_22.4A	2.56	6.49	4,181	2,582	20.27	148.44	34.4	40.7	0.18	0.19
RK_30_22.5B	2.56	6.41	4,236	2,628	23.43	150.71	33.6	42.1	0.09	0.16
Mean	2.49	8.44	4,106	2,510	17.39	124.62	30.6	37.5	0.2	0.18
Standard deviation	0.09	3.23	221	133	5.512	37.01	5.5	5.1	0.06	0.03

Siratovich et al.

Siratovich et al. Geothermal Energy 2014 2:10 doi:10.1186/s40517-014-0010-4

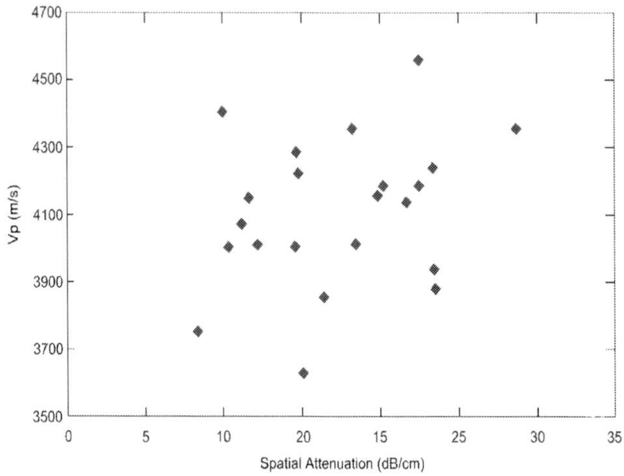

Figure 9: Spatial attenuation (αs) of axial compressional P-wave velocity of Rotokawa Andesite. Attenuation was calculated by the method suggested by Martínez-Martínez et al. ([2011]) utilizing transmission of the ultrasonic wave and maximum attenuation of the waveform plotted versus axial compressional wave velocity (Vp) obtained under an axial stress of 10 MPa.

Uniaxial Compressive Strength and Static Elastic Moduli

In order to characterize the mechanical behavior of the Rotokawa Andesite, the 22 samples of Table 3 were loaded in compression to failure. The dataset shows a large range of UCS (as observed for the other physical properties), from 60 to 211 MPa. The stress-strain behavior of the andesites is very similar across the range of strengths, as shown in Figure 10, which reports curves that best represent the dataset and behavior of the Rotokawa Andesite under uniaxial compression. All specimens in the dataset show brittle behavior as evidenced by the stress-strain relationships and bolstered by analysis of the AE activity. An increase of AE between dilatancy (σ_{cd}) and failure is a benchmark of brittle failure (e.g., Brace and Bombolakis [1963]; Rutter[1986]; Ashby and Sammis

[1990]; Heap and Faulkner [2008]), as seen in Figure 10. Weaker specimens showed lower overall AE energy output than higher strength specimens. Static Young's moduli range between 19.9 and 43.7 GPa and static Poisson's ratio between 0.09 and 0.34. Our data shows that as porosity and crack surface area increases, the UCS of the rock decreases (Figure 11A,B). Further, the UCS increases as axial P-wave velocity increases (Figure 11C).

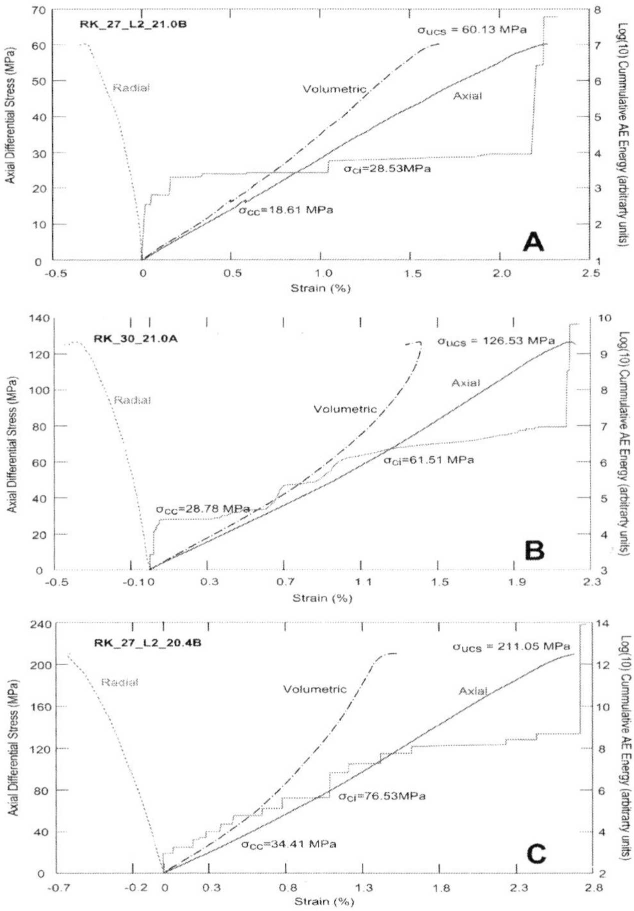

Figure 10: Stress–strain behavior of the Rotokawa Andesite.Samples were subject to constant strain rate loading (1×10^{-5}/s) and monitored for associated arbitrary acoustic emission energy output. All samples in this study

display brittle failure. (A) Samples with low UCS generally develop a single fracture plane. (B) Samples near mean UCS develop several fracture planes. (C) Samples with very high UCS showed explosive, catastrophic failure into several large and small pieces with no distinct failure plane.

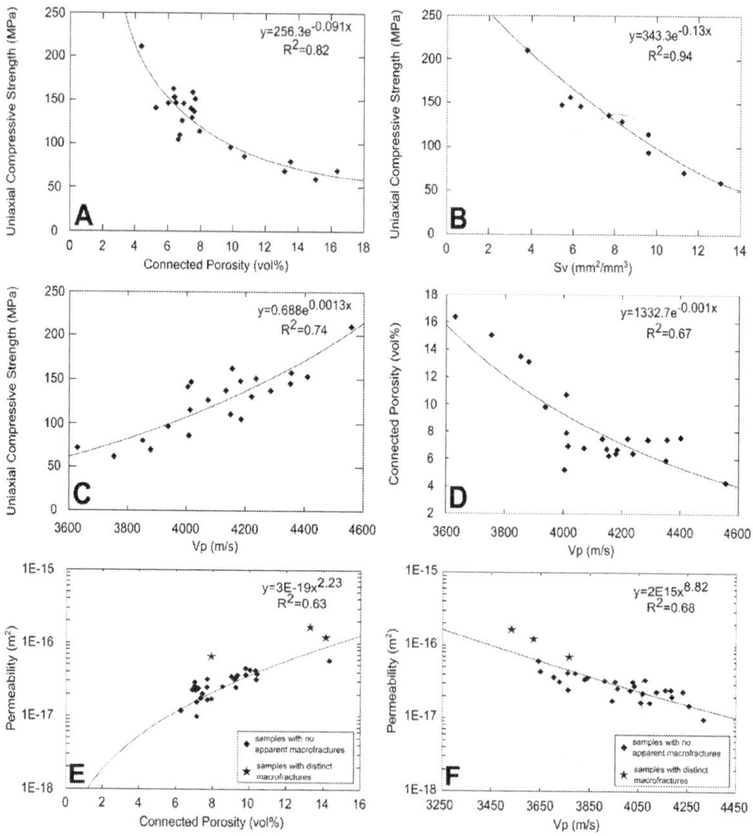

Figure 11: Key empirical relationships of the Rotokawa Andesite. (A) Connected porosity (vol%) plotted versus uniaxial compressive strength (MPa) for Rotokawa Andesite at ambient pressures and temperatures. (B) Crack area per unit volume (Sv) as measured from reflected light thin sections (method recommended by Underwood [1970]) plotted versus uniaxial compressive strength (UCS) for the Rotokawa Andesite. (C) Axial P-wave velocity (Vp) as measured at ambient temperatures under an axial load of 10 MPa versus uniaxial compressive strength values for the Roto-

kawa Andesite at ambient temperature. (D) Axial compressional wave velocity (Vp) plotted versus connected porosity (vol%) as measured by the dual weight method (Ulusay and Hudson [2007]). (E) Semi-log plot of connected porosity measured by the triple-weight method (Ulusay and Hudson [2007]) versus argon gas permeability with effective pressure of 0.5 MPa. (F) Semi-log plot of axial compressional wave velocity (Vp) plotted versus argon gas permeability.

Permeability

Our argon permeability measurements show that for the tested samples, permeability ranges from 9.82×10^{-18} m^2 to 1.66×10^{-16} m^2 (Table 4). The results show a trend of increasing permeability with increasing porosity (Figure 11E). We observe that three of the samples contain macrofractures (black stars on Figure 11E,F) and have higher permeabilities that slightly deviate from the trend of the dataset. We also note that as permeability increases, the axial compressional wave velocity decreases (Figure 11F); further, those samples with distinct macrofractures show lower compressional velocities and higher permeabilities when compared with samples of similar porosity.

Table 4: Results of density, porosity, argon permeability, and acoustic velocity measurements for Rotokawa Andesite

Sample source: well number, depth, name	Bulk dry density (g/cm3)	Connected porosity (vol%)	Argon permeability (m2)	Axial P-wave velocity(m/s)
RK_27_L2_2120.4A_1	2.55	7.24	1.78E−17	3,943
RK_27_L2_2120.4A_2	2.55	7.06	9.82E−18	4,318
RK_27_L2_2121.1A	2.34	13.97	1.25E−16	3,621
RK_27_L2_2123.3A_1	2.48	9.22	3.32E−17	3,911
RK_27_L2_2123.3A_2	2.46	10.25	4.13E−17	3,793
RK_27_L2_2123.7A_1	2.52	9.73	3.69E−17	3,840
RK_27_L2_2123.7A_2	2.50	10.35	3.72E−17	3,704
RK_27_L2_2124.1B_1	2.54	7.63	3.24E−17	3,954

RK_27_L2_2124.1B_2	2.56	6.93	2.81E−17	4,032
RK_27_L2_2121.0A_1	2.42	13.14	1.66E−16	3,532
RK_27_L2_2121.0A_2	2.33	14.23	6.09E−17	3,642
RK_28_2310.3A_1	2.46	9.94	4.26E−17	3,760
RK_28_2310.3A_2	2.48	9.70	4.44E−17	3,650
RK_28_2310.8A_1	2.50	8.95	3.45E−17	3,830
RK_28_2310.8A_2	2.50	9.22	3.43E−17	4,076
RK_28_2310.9C_1	2.46	10.30	3.21E−17	3,725
RK_28_2310.9C_2	2.53	7.64	1.65E−17	4,096
RK_28_2311.1B_1	2.55	7.04	1.53E−17	4,259
RK_28_2311.1B_2	2.56	7.35	2.05E−17	4,190
RK_28_2311.3B_1	2.50	9.10	3.21E−17	4,026
RK_28_2311.3B_2	2.49	9.17	2.50E−17	4,183
RK_30_2320.6A_1	2.53	7.90	7.10E−17	3,765
RK_30_2320.6A_2	2.56	7.79	1.70E−17	4,061
RK_30_2321.0B_1	2.50	8.24	2.50E−17	3,762
RK_30_2321.0B_2	2.54	8.47	2.50E−17	4,164
RK_30_2321.2B_1	2.58	7.11	2.38E−17	4,235
RK_30_2321.2B_2	2.55	6.98	2.37E−17	4,122
RK_30_2321.5A	2.57	6.92	2.60E−17	3,965
RK_30_2322.3A_1	2.54	7.61	2.44E−17	4,016
RK_30_2322.3A_2	2.57	6.83	2.27E−17	4,069

Siratovich et al.

Siratovich et al. Geothermal Energy 2014 2:10 doi:10.1186/s40517-014-0010-4

DISCUSSION

Micromechanical Interpretation

We have shown that the Rotokawa Andesite contains a pervasive network of isotropic microcracks. Due to their isotropic distribution, the majority of these microcracks are consistent with the results of thermal stressing (Fredrich and Wong [1986]; Reuschlé et al. [2006]; Wang et al. [1989]; David et al. [1999]; Heap et al. [2014]). Indeed, the Rotokawa Andesite has experienced several cycles of heating and cooling: the initial eruption of the andesite, burial in a faulted graben, hydrothermal alteration, and the eventual exhumation during core recovery (Rae [2007]; Lim et al. [2012]). Our microstructural analysis has highlighted that the pervasive microcracking appears independent of lithology, original mineralogy, and secondary (hydrothermal alteration) mineralogy.

The intense microcracking in our samples has shown to be a significant factor in all of the measured physical properties. First, microcracking has greatly reduced the propagation velocity of elastic waves through the andesite. We see a clear correlation of crack area per unit volume (Sv) to the observed compressional wave velocities (Figure 8D) and interpret this to be attenuation of the compressional wave through the cracked intracrystalline and intercrystalline boundaries that are abundant in the andesite (e.g., Figures 3 and 4). Many authors (e.g., Vinciguerra et al. [2005]; Keshavarz et al. [2010]; Blake et al. [2012]; Heap et al. [2014]) have also shown that the elastic wave velocities can be highly attenuated by the presence of microcracks.

Second, the crack surface area and UCS have yielded an excellent correlation (Figure 11B). As noted by Walsh ([1965a], [b]), David et al. ([1999]), and Chaki et al. ([2008]), the density of the cracks within a specimen is critical in dictating its strength. The development of microcracks during uniaxial compression, and the coalescence of these cracks (newly formed and pre-existing), leads to the failure of the sample (Brace et al. [1966]; Bieniawski [1967]). In samples that already show relatively high crack densities, less energy is required to coalesce existing cracks and thus they are inherently weaker (David et al. [1999]; Ferrero and Marini [2001]; Keshavarz et al. [2010]). By utilizing AE monitoring during our UCS testing, we observe that fewer events occur during uniaxial compression in weaker samples than those with higher strength (Figure 10), indicating that there are far more pre-existing cracks in the weaker samples (Hardy [1981]; Eberhardt et al.[1998]; Nicksiar and Martin [2012]). Thus, the presence of pre-existing microcracks in the Rotokawa Andesite is shown to exert a strong control on their uniaxial compressive strength.

Permeability is one of the most important properties of a geothermal system. In this study, we have seen that porosity (and bulk sample density) and strength are related to the extent of the microcracking in the andesite. We did not measure the crack surface area in the samples used for our permeability measurements (the samples will be used for future studies; calculating crack surface area required destructive thin section preparation). However, we can, by proxy, assume a correlation between permeability and the extent of the microfracture network. We show that there is a clear inverse relationship between the sample's permeability and P-wave velocity such that as permeability increases, compressional wave velocity decreases (Figure 11F). These results are consistent with the many investigations have shown a clear link between reduced elastic wave velocities and increased permeability (David et al. [1999]; Vinciguerra et al. [2005]; Chaki et al.[2008]; Nara et al. [2011]; Faoro et al. [2013]; Heap et al. [2014]). While we have not measured the relationship of crack density to permeability directly in our dataset, we show that Sv and Vp are inversely related (Figure 8D), and a similar relationship exists between Vp and permeability. Therefore, we can infer that those samples with higher crack surface areas will be inherently more permeable.

Key Empirical Relationships

In this section, we present relationships of singular variables that could be readily and easily measured either using photomicrography or geophysical logging tools and their correlation to more complicated and pertinent physical properties. All of these parameters are singularly measurable variables that do not rely on complex formulae for their derivation (such as dynamic Young's Modulus or Poisson's ratio) and so have been selected to be the key relationships that we present with relevance to the Rotokawa Andesite.

Porosity and UCS

An exponential correlation between sample porosity and UCS exists (Figure 11A). Such correlations have been utilized by several authors (e.g., Vernik et al. [1993]; Li and Aubertin[2003]; Palchik and Hatzor [2002]; Kahraman et al. [2005]; Chang et al. [2006]; Palchik [2013]; Pola et al. [2014]) for a variety of clastic and volcanic rocks and concrete materials. These authors present empirical fits for the correlation of physical properties versus UCS and show a wide range of correlation within their respective datasets with R^2 values from near 0.6 to as high as 0.95. We propose that our empirical fit between porosity and UCS (an exponential fit with a correlation factor of 0.82, Figure 11A) can provide useful estimations of the strength of the reservoir rocks within the Rotokawa Andesite reservoir. By utilizing estimations of UCS derived from the correlation of porosity, the minimum strength of the rocks can be applied to important engineering issues such as wellbore stability (Chang et al. [2006]; Schöpfer et al. [2009]).

Vp and UCS

There is an exponential correlation between strength and Vp with an R^2 value of 0.74 (Figure 11C). As noted by Kahraman ([2001]), the relationship between Vp and UCS is generally nonlinear and the higher the strength of the material, the more scattered the data points. Heap et al. ([2014]) came to similar conclusions following measurements on andesitic rocks from Volcán de Colima (Mexico). In our study, there is an increasing trend of strength with increasing Vp but, as shown in Figure 9, there is a high degree of spatial anisotropy with respect to Vp such that a robust correlation of strength to elastic wave velocity is difficult to obtain. However, Vp is a widely utilized logging tool in borehole geophysics (Chang et al. [2006]), and using the correlation that we have obtained, a minimum strength criteria could be established from the response of the logging tool. This is an important correlation as geophysical logging is much easier, faster, and more efficient than cutting spot cores (as the core for this study was obtained), and so the development of empirical correlations to constrain strength such as that seen in Figure 11B can help mitigate risk and reduce the cost associated with geothermal drilling programs.

Vp and Porosity

Correlations between Vp and porosity show an increasing trend of porosity with decreasing Vp (Figure 11D, also observed by Al-Harthi et al. [1999]; Rajabzadeh et al. [2011]; Tugrul and Gurpinar [1997]; Heap et al. [2014]). This can be attributed to both the pore structure distribution and the degree of microcracking within the andesites. It is clear from microstructural analysis (using both optical and scanning electron microscope analyses) that a large proportion of the porosity in the Rotokawa Andesite is likely to be composed of (macro- and mesoscale) fractures and microcracks (e.g., Figures 6 and 7).

An explanation for the variation and wide distribution of the elastic wave velocity data for samples with similar porosities (specifically with regard to those data that range from 4,000 to 4,400 m/s) is that there must be a variable pore (vug/vesicle) content or hydrothermal alteration between the samples. The presence of pores will greatly augment the porosity (due to their aspect ratio) but will have comparatively little influence, compared to the microcracks, on the P-wave velocity. The application of our exponential relationship (Figure 11D) can give a rough approximation for seismic velocities derived from connected porosity, or vice versa. This may be useful during the drilling of additional wells at Rotokawa where porosity can be measured at the wellsite and yield a rough approximation for P-wave velocities and, as such, tie back to our empirical correlations of strength (Figure 11C).

Permeability and Porosity

Our permeability and porosity data show that there is a clear trend of increasing porosity with increased permeability for the Rotokawa Andesite (Figure 11E), a common observation in multiple lithologies (e.g., Heard and Page [1982]; Géraud [1994]; Stimac et al. [2004]; Chaki et al. [2008]; Watanabe et al. [2008]; Heap et al. [2014]). We observe that our relationship between porosity and permeability can be described by a power law correlation and is consistent with the Kozeny-Carman relation (Guéguen and Palciauskas [1994], see the 'Application of micromechanical and geometrical permeability models' section). The dependence of permeability on porosity is generally explained by the assumption that a more connected pore space (cracks and pores) provides more efficient pathways for fluid migration (e.g., Costa [2006]; Chaki et al. [2008]). We do however need to consider those data points that have a very similar value of permeability (approximately 3.2×10^{-17} m^2, Table 4), with a porosity range of 7.6 to 10.3 vol% that indicate that there is variability of the samples with respect to permeability that may be reflected in the tortuosity of the porous network. This is consistent with the findings of Bernard et al. ([2007]) and Heap et al. ([2014]) such that the permeability in volcanic rocks is highly dependent upon connectivity of the microstructure.

With respect to microstructure, we have shown that the porosity is very closely linked to crack surface area (Figure 8D) and, thus, that increasing crack density corresponds to a sample with a higher permeability. The three samples that lie slightly outside the trend of the dataset display distinct mesofractures (black stars in Figure 11E,F) and that these mesofractures greatly enhance the permeability of the samples without significantly increasing their porosity. These specimens show higher than average permeability for their porosity, which supports the conclusions of Stimac et al. ([2008]) that meso- and macrofractures are critical in controlling the permeability of geothermal reservoir systems. On the large scale, macrofractures are necessary for fluid production from geothermal reservoirs, but the microstructural characteristics of the host rocks cannot be neglected when considering fluid flow, storage capacity, and total permeability of the reservoir (Jafari and Babadagli [2011]).

The robust relationship between porosity and permeability has wider-scale reservoir applications where the need to understand reservoir rock permeability (the mass itself, not those portions with highly macroscopic fractures e.g., Massiot et al. [2012]) is important for reservoir forecasting and modeling. Measurements of porosity can then yield a good approximation of the permeability of the intact reservoir rock at Rotokawa through our power law correlation (Figure 11E). However, we urge caution if the porosity falls outside our measured range. As porosity is a readily measureable property by geophysical logging tools (Ellis and Singer [2008]), the response from such a tool, together with our empirical fit, can give engineers and geoscientists an approximation of the matrix permeabilities in the Rotokawa Andesite.

Permeability and Acoustic Velocities

There is a clear inverse relationship between our measurements of permeability and P-wave velocity (Figure 11F) such that the more permeable the sample, the slower the compressional wave velocity. These findings are consistent with the findings of many other authors (e.g., Vinciguerra et al. [2005]; Chaki et al. [2008]; Nara et al. [2011]; Heap et al. [2014]). The correlation of such properties is an excellent tool for understanding the micro- and mesoscopic fracture networks and their relation to permeability in the Rotokawa Andesite as follows: (1) we have shown that the porosity and crack density are closely linked (Figure 8A), (2) acoustic velocity and crack density are closely linked (Figure 8D), and (3) there is a power law correlation of Vp and permeability (Figure 11F). Thus, there is a direct link of P-wave velocity to permeability that is reliant on the crack densities of the samples. The relationship we present in Figure 11F shows a power-law fit which would indicate that the hydraulic radii of the pore space (pore and cracks) are similar in size but that the higher the concentration of cracks, the higher the permeability we observe (Bourbie and Zinszner [1985]).

Similarly, there are occasional mesofractures (with apertures less than 1-mm width; we note that these fractures are much smaller than those described in Massiot et al. [2012]) in the samples that deviate from the rest of the dataset (black stars, Figure 11F). The presence of these macrofractures increases permeability (by a factor of 2) and also appears deleterious to elastic wave propagation (all the three samples containing mesofractures have low elastic wave velocities, although we cannot separate the influence of meso- and microcracks on the velocities of these samples). Further, elastic waves are useful for the detection of cracks in rock and concrete (Chaki et al. [2008]; Heap et al. [2013]), and a decreased elastic wave velocity correlates well to more permeable media which is observed by the three outlying, higher permeability, lower elastic wave velocity samples.

The correlation between elastic wave velocity and permeability outside the laboratory has potentially far-reaching value for the prediction of reservoir permeability interactions from wireline logging and larger-scale seismic and microseismic surveys. There is a complex microseismic network installed at Rotokawa, and the location of earthquake activity has been closely linked to macroscopic permeability within the reservoir (Sewell et al. [2013]; Sherburn et al. [2013]). The existing model of the velocity structure at depth could then be further refined using our acoustic velocity and permeability data for reservoir rock matrix. This may allow a deeper and more accurate understanding of the distribution of permeability at depth.

Additionally, the data we have presented can also be used to infer values of matrix permeability from acoustic wireline logs (dipole sonic) used during exploration at nearby Ngatamariki Geothermal Field (Wallis et al. [2009]). Should similar geophysical logging be used in future wells drilled at Rotokawa, the matrix permeability may be estimated using the relationship we present here. In addition, the coupling of these data with microseismic data could allow a significant increase in understanding the complexity of the Rotokawa Andesite reservoir. While we are aware that macrofractures augment the elastic wave velocity during routine acoustic profiling (e.g., Barton and Zoback [1992]), our laboratory data show that although samples containing mesofractures (i.e., on the sample scale) are shifted to higher permeabilities and elastic wave velocities, they do not stray too far away from the trend extrapolated from our power-law relationship. Despite this, we urge a certain degree of caution, based on the potential presence of large-scale fractures, when estimating permeability using our derived permeability-elastic wave velocity relationship.

Application of Micromechanical and Geometrical Permeability Models

Extracting empirical relationships between laboratory-derived rock properties is useful; however, the parameters are not easily related to independently measurable quantities (i.e., they lack a physical basis). Micromechanical (e.g., the wing-crack model of Ashby and Sammis [1990]) and geometrical permeability models (e.g., the Kozeny-Carman relation, Guéguen and Palciauskas[1994]) can be better constrained as the parameters used in such models have a clear physical meaning. In this section, we attempt both sliding wing-crack modeling and Kozeny-Carman permeability modeling to investigate the microstructural controls on deformation and fluid flow, respectively.

Micromechanical Modeling

Micromechanical modeling can provide useful insights in the mechanics of compressive failure in brittle rock (Wong and Baud [2012]). Since the rocks of this study contain high microcrack densities, we will use the sliding wing-crack model of Ashby and Sammis ([1990]). This model idealizes the rock microstructure as an elastic continuum embedded with inclined (45°) microcracks (of length 2c). These microcracks act as stress concentrators for the initiation of 'wing' cracks when the frictional resistance of the closed crack is overcome and the stress at the tip of the crack exceeds the critical stress intensity factor (K_{IC}). The cracks can then propagate in the direction of the maximum principal stress. Eventually, the cracks coalesce, resulting in the failure of the elastic medium. In the case of uniaxial compression, Baud et al. ([2014]) derived an analytical approximation to estimate UCS:

$$\mathrm{UCS} = \frac{1.346}{\sqrt{1+\mu^2}-\mu} \frac{K_{\mathrm{Ic}}}{\sqrt{\pi c}} D_0^{-0.256} \tag{10}$$

where μ is the friction coefficient of the sliding crack and D_0 is an initial damage parameter that is a function of the angle of the initial microcrack with respect to the maximum principal stress and the initial number of sliding cracks per unit area (Ashby and Sammis [1990]).

The analytical solution (that assumes an initial crack angle of 45°) presented above contains five parameters. We have, through experimental data and observations, a good handle on three of the parameters: (1) we have measured the UCS of 22 samples (Table 3), (2) μ rarely deviates from 0.6 to 0.7 (Byerlee [1978]), and (3) c can be determined from optical microscopy (we determinedc by measuring the approximate average length of the microcracks under the microscope). We do not have a laboratory-determined value for K_{IC}. While the K_{IC} of andesite has been previously measured to be about 1.5 $MPam^{0.5}$ (Ouchterlony [1990]; Obara et al. [1992]; Tutluoglu and Keles[2011]; Nara et al. [2012]), there is no guarantee that this value is representative of the Rotokawa Andesite, which is likely to be lower than these values due to hydrothermal alteration. We therefore have chosen a slightly lower K_{IC} of 1.0 $MPam^{0.5}$ for our analysis. Using our UCS data, we can solve Equation 10 to assign a value of D_0 to each experiment (using $\mu = 0.6$; $K_{IC} = 1.0$; c = 0.001 m). The goal of such analysis, assuming that the other parameters remain roughly constant between different samples/cores, is to estimate D_0 using an easily measured physical property, such as Vp (therefore allowing us to predict rock strength, using the micromechanical model, from Vp measurements alone). Our analysis shows that D_0 ranges from 0.0019 to 0.26 for the 22 measured samples (with average of 0.039). D_0 is plotted against the crack area per unit volume (Sv) and Vp in Figure 12 and indicates that D_0 increases as Sv increases (Figure 12A). While this may appear logical (D_0 is a function of the initial crack density), it serves as an encouraging proof of the concept. The increase in D_0 with crack density is not linear; D_0 increases more rapidly beyond 10 mm^{-1} (Figure 12A). We also see that Vp decreases with increasing D_0; in detail, Vp decreases rapidly as D_0 increases from 0 to 0.05 and then decreases more gradually above 0.05. Unfortunately, the relationship between D_0 and Vp is a little more clouded (the data are more scattered, Figure 12B) and probably represents variable vesicle density (the model assumes that vesicles do not play a role in failure in compression) and hydrothermal alteration (we assume thatK_{IC} and the average crack lengths are constant). The conclusion of this pilot analysis is that the variability within the Rotokawa Andesite is potentially too large to permit meaningful microstructural wing-crack modeling, but greater success could be achieved with laboratory-determined values for K_{IC}. Therefore, if micromechanical modeling is to be deployed as a feasible method to predict the strength of Rotokawa Andesite reservoir rocks, the samples/cores should be grouped by their alteration, and K_{IC} measured for each alteration group.

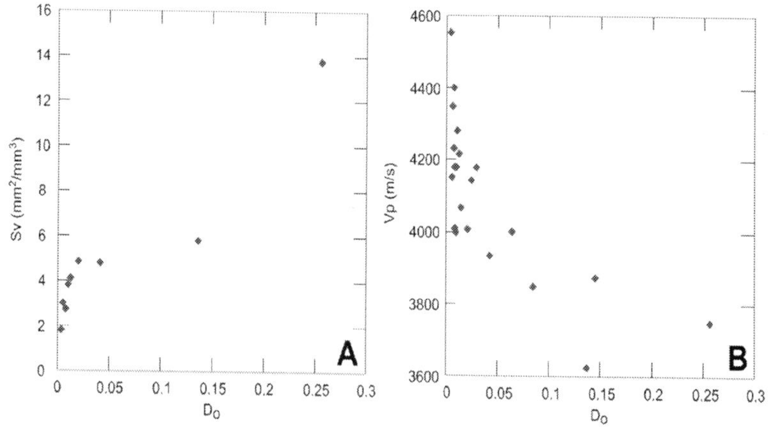

Figure 12: Results of geometric modeling for Rotokawa Andesite. (A) Initial damage parameter D_0 as predicted by Equation 10 and described by Baud et al. ([2014]) plotted versus calculated crack densities by the meth-

od of (Underwood [1970]). (B) Prediction of compressional wave velocity (Vp) as a function of the initial damage parameter D_0 the relationship between D_0 and Vp shows a moderate correlation with high initial damage parameter but becomes quite clouded in those samples with a very small calculated D_0 (see text for further expansion on this relationship).

Permeability Modeling

Kozeny-Carman models are those that use the notion of a hydraulic radius (see Guéguen and Palciauskas [1994]) to correlate porosity and permeability. Forms of the Kozeny-Carman relation have previously been used in the study of volcanic rocks (e.g., Saar and Manga [1999]; Costa[2006]; Bernard et al. [2007]; Heap et al. [2014]), while others have used a heavily simplified version (e.g., Rust et al. [2003]; Mueller et al. [2005]; Lavallée et al. [2013]). The Kozeny-Carman relation is of the form:

$$k_{\mathrm{KC}} = \frac{\varphi (r_{\mathrm{H}})^2}{b\tau^2} \tag{11}$$

where k_{KC} is the permeability, φ is the connected porosity, b is a geometrical factor, τ is the tortuosity of the equivalent channel (i.e., the ratio of its actual to nominal length), and r_H is the hydraulic radius (i.e., the volume of pores divided by the surface of the pores). The power law exponent for our data (excluding those samples with macrofractures) is about 2.2 (Figure 11E) and is therefore consistent with the Kozeny-Carman model (Bourbie and Zinszner [1985]; Doyen[1988]). In detail, one would expect a power law exponent of 2 or 3 if the elements controlling the permeability are tubes or cracks, respectively (Guéguen and Palciauskas [1994]). Our power law exponent is between these two values. This is somewhat surprising, considering the pervasive fracture network in these materials, but could reflect flow through a combination of cracks and tubes or our limited porosity range. Since the entire dataset can be described by a single power law exponent, we conclude that within our limited range of connected porosities, there is no dramatic shift in pore space connectivity or tortuosity, as was the case for Fontainebleau sandstone at a porosity of 9 vol% (Bourbie and Zinszner [1985]) and andesite samples from Volcán de Colima (Mexico) at a porosity of about 11 vol% (Heap et al. [2014]). Extrapolating to porosities outside this range may be treacherous especially to lower porosities where samples may become subject to a higher power law exponent. However, within the dataset, the model predicts an increase in permeability of a factor of 1.5 for an increase in porosity of 1 vol% (an increase not uncommon for rock following a thermal stressing episode; e.g., Chaki et al. [2008]).

Application of Results to Geothermal Exploration and Utilization

The relationships between porosity, acoustic wave velocities, strength, and permeability are valuable for understanding a geothermal reservoir. Our data indicate strong correlations between these parameters, as observed by Stimac et al. ([2004], [2008]) amongst others. The data we have obtained are from cores sourced from three production wells. Such materials are very expensive to obtain, time consuming, and, if coring did not go as planned, can pose great risk of losing the well (Finger and Blankenship [2010]; Hole [2013]). The microstructural and empirical correlations presented in this study can be applied to new wells drilled in geothermal environments and can help refine studies on pre-existing wells, if our correlations hold true at the reservoir scale. Some physical parameters, such as porosity and elastic wave velocities, are easily obtainable through the use of down-hole geophysical logging suites. The empirical correlations shown in this study (bolstered by our application of classical models) show that readily measurable physical properties may therefore be used to predict more complex and pertinent properties such as strength and permeability. Such correlations and calibrations are common in the hydrocarbon industry especially during exploration drilling (e.g., Vernik et al. [1993] and references therein), and we consider that our dataset can help improve the understanding of the Rotokawa reservoir while minimizing the risk to future drilling operations.

A clear understanding of the factors that control reservoir rock permeability is fundamental for the planning of stimulation and enhancement operations that may be necessary as the Rotokawa field and reservoir dynamics change with continued production. The need to drill additional wells or re-work pre-existing wells may become apparent and the ease at which the reservoir can accept and deliver fluids (i.e., its permeability) will be of utmost importance. The thermal stimulation of injection wells has taken place at Rotokawa for some time by the injection of power-plant condensates and spent brines (Siega et al. [2009]), but the technique may play a significant role in enhancing production wells at some future stage.

Therefore, a deeper understanding of how permeability may be increased through stimulation is important. The application of models such as the Kozeny-Carman may provide insight to permeability enhancement. An increase in the porosity of reservoir rock by 1 vol%, according to the geometrical model, should increase the permeability by a factor of 1.5. In the case of an aging field and aging wellbores, such an increase could greatly extend the life of the field. In the interests of keeping geothermal projects commercially economic, the fundamental understanding of the reservoir rock properties become essential to the continued utilization and management of the field.

CONCLUSIONS

Our study provides a comprehensive evaluation of the physical and mechanical properties of the Rotokawa Andesite through a multi-disciplinary approach. We have evaluated the Rotokawa Andesite from the microstructural to macroscopic scale and have presented robust datasets that permit the correlation and comparison of important physical properties to geothermal exploitation. A comprehensive understanding of how the relationships of microstructural texture influence key physical properties such as strength and permeability, essential for the optimal utilization of a geothermal resource have been investigated.

Further, we summarize our conclusions as follows:

1. We have shown that the presence and intensity of microfracturing in the Rotokawa Andesite are the predominant controlling factors on physical and mechanical properties. The behavior of these properties is also shown to be largely independent of the alteration mineralogy as we see similar alteration intensities in the samples we have studied.
2. Guided by a systematic understanding of role of microfractures, we show that empirical correlations of strength and porosity can be developed and applied to field scale engineering problems. We have shown that as the porosity increases, the strength decreases and elastic wave velocities are attenuated. Similarly, we show that permeability increases with increased porosity and reduced acoustic velocity. These findings are applicable if geophysical logging tools be used after the drilling of wells to ascertain properties such as porosity; our dataset provides useful means to address complex reservoir problems.
3. We further boost our empirical correlations by applying classical physical models based on sound physical theory to predict both UCS and permeability through understanding of the microstructure. We have applied these models with some success, but these models are best-suited for homogeneous, isotropic materials. Further work to constrain these models should include laboratory investigations of fracture toughness (K_{IC}) and the factors that influence this variable. However, our fit for the damage criterion D_0 is acceptable and builds the foundations for future understanding and may permit the construction of similar better constrained models.
4. The study comprises a large dataset with a goal to further push the knowledge that can be sourced from a geothermal environment such as the Rotokawa Andesite. The properties that we have evaluated are very difficult to constrain without direct information from rocks sourced from the reservoir. Geothermal reservoirs are complex, and harsh environments from which the recovery of intact core can present a

significant and financially risky challenge. The results that we present here help us to understand this complex reservoir environment by their application to field scale engineering and geological issues.

5. Our analyses have provided quantifiable and measurable physical properties of the Rotokawa Andesite. However, the dataset is not exhaustive. Further studies need to be carried out to replicate near-reservoir conditions in the laboratory and should focus on permeability at the high confining pressures and temperatures found in the reservoir. Additionally, mechanical testing such as triaxial, tensile strength, and fracture toughness experiments should be conducted under high-temperature conditions, potentially in the presence of reservoir-type fluids to aid in predictions of reservoir behavior and geomechanical modeling under conditions as close as possible to those found in the reservoir.

AUTHORS' CONTRIBUTIONS

PS performed the majority of the experimental work and composition of the manuscript. MH assisted in revision and suggestions for the manuscript and performed the permeability and micromechanical modeling. MV provided assistance in experimental work and processing of mechanical data. JC provided revision and petrological input. TR provided the permeability measurements. All authors read and approved the final manuscript.

ACKNOWLEDGEMENTS

The authors wish to thank Mighty River Power Company Ltd. for a generous grant for PS, which allowed collaboration with MH and TR. We also wish to thank the Rotokawa Joint Venture, a joint venture between the Tauhara North No. 2 Trust and Mighty River Power Company Ltd. for the core material used in this study. The staff of the Department of Geological Sciences at the University of Canterbury were invaluable in assisting in all aspects of this research. The Brian Mason Trust also provided for the transportation and delivery of the core to UC. The authors of this study also acknowledge a Hubert Curien Partnership (PHC) Dumont d'Urville grant (grant number 31950RK) which has assisted the France-New Zealand collaboration for this and future projects. MH was partly funded by the LABEX ANR-11-LABX-0050_G-EAU-THERMIE-PROFONDE framework (funding from the state managed by the French National Research Agency as part of the Investments for the Future Program).

REFERENCES

1. Al-Harthi AA, Al-Amri RM, Shehata WM (1999) The porosity and engineering properties of vesicular basalt in Saudi Arabia. Eng Geol 54:313-320
2. Ashby MF, Sammis CG (1990) The damage mechanics of brittle solids in compression. Pure Appl Geophys 133:489-521
3. Barton CA, Zoback MD (1992) Self-similar distribution and properties of macroscopic fractures at depth in crystalline rock in the Cajon Pass Scientific Drill Hole. J Geophys Res 97(B4):5181-5200
4. Baud P, Wong T-F, Zhu W (2014) Effects of porosity and crack density on the compressive strength of rocks. Int J Rock Mech Min Sci : (in press)
5. Bernard M-L, Zamora M, Géraud Y, Boudon G (2007) Transport properties of pyroclastic rocks from Montagne Pelée volcano (Martinique, Lesser Antilles). J Geophys Res 112:
6. Bibby HM, Caldwell TG, Davey F, Webb T (1995) Geophysical evidence on the structure of the Taupo Volcanic Zone and its hydrothermal circulation. J Volcanol Geotherm Res 68:29-58
7. Bieniawski ZT (1967) Mechanism of brittle fracture of rock part II - experimental studies. Int J Rock Mech Min Sci 4:407-423

8. Blake OO, Faulkner DR, Rietbrock A (2012) The effect of varying damage history in crystalline rocks on the P- and S-wave velocity under hydrostatic confining pressure. Pure Appl Geophys 170:493-505
9. Bloomberg S, Rissmann C, Mazot A, Oze C, Horton T, Kennedy B, Werner C, Christenson B, Pawson J (2012) Soil gas flux exploration at the Rotokawa Geothermal Field and White Island, New Zealand. In: Proceedings, Thirty Sixth Workshop on Geothermal Reservoir Engineering. Stanford University, Stanford, California.30 January 30 to 1 February 2012
10. BourbieT, Zinszner B (1985) Hydraulic and acoustic properties as a function of porosity in Fontainebleau sandstone. J Geophys Res 90:11524-11532
11. Brace WF, Bombolakis EG (1963) A note on brittle crack growth in compression. J Geophys Res 68:3709-3713
12. Brace WF, Paulding B, Scholz C (1966) Dilatancy in the fracture of crystalline rocks. J Geophys Res 71:3939-3953
13. Brace WF, Walsh JB, Frangos WT (1968) Permeability of granite under high pressure. J Geophys Res 73:2225-2236
14. Byerlee JD (1978) Friction of rocks. Pure Appl Geophys 116:615-626
15. Chaki S, Takarli M, Agbodjan WP (2008) Influence of thermal damage on physical properties of a granite rock: porosity, permeability and ultrasonic wave evolutions. Constr Build Mater 22:1456-1461
16. Chang C, Zoback MD, Khaksar A (2006) Empirical relations between rock strength and physical properties in sedimentary rocks. J Pet Sci Eng 51:223-237
17. Cole JW (1990) Structural control and origin of volcanism in the Taupo volcanic zone, New Zealand. Bull Volcanol 52:445-459
18. Collar RJ, Browne PRL (1985) Hydrothermal eruptions at The Rotokawa Geothermal Field, Taupo Volcanic Zone, New Zealand. In: Proceedings of the seventh New Zealand

geothermal workshop, University of Auckland. Geothermal Institute, Auckland, New Zealand 6–8 November 1985

19. Costa A (2006) Permeability-porosity relationship: a re-examination of the Kozeny-Carman equation based on a fractal pore-space geometry assumption. Geophys Res Lett 33:
20. David C, Menendez B, Darot M (1999) Influence of stress-induced and thermal cracking on physical properties and microstructure of La Peyratte granite. Int J Rock Mech Min Sci 36:433-448
21. Diamantis K, Gartzos E, Migiros G (2009) Study on uniaxial compressive strength, point load strength index, dynamic and physical properties of serpentinites from Central Greece: test results and empirical relations. Eng Geol 108:199-207
22. Diederichs M, Kaiser P, Eberhardt E (2004) Damage initiation and propagation in hard rock during tunneling and the influence of near-face stress rotation. Int J Rock Mech Min Sci 41:785-812
23. DiPippo R (2008) Geothermal power plants: principles, applications, case studies and environmental impact. Elsevier Ltd, Oxford.
24. Doyen PM (1988) Permeability, conductivity, and pore geometry of sandstone. J Geophys Res 93:7729-7740
25. Eberhardt E, Stead D, Stimpson B, Read RS (1998) Identifying crack initiation and propagation thresholds in brittle rock. Can Geotech J 35:222-233
26. Ellis DV, Singer JM (2008) Well logging for earth scientists. Springer, Dordrecht.
27. Faoro I, Vinciguerra S, Marone C, Elsworth D, Schubnel A (2013) Linking permeability to crack density evolution in thermally stressed rocks under cyclic loading. Geophys Res Lett 40:2590-2595
28. Ferrero AM, Marini P (2001) Experimental studies on the mechanical behaviour of two thermal cracked marbles. Rock Mech Rock Eng 34:57-66

29. Finger J, Blankenship D (2010) Handbook of best practices for geothermal drilling. Sandia National Laboratories, Albuquerque.
30. Fredrich JT, Wong T (1986) Micromechanics of thermally induced cracking in three crustal rocks. J Geophys Res 91:12743-12764
31. Géraud Y (1994) Variations of connected porosity and inferred permeability in a thermally cracked granite. Geophys Res Lett 21:979-982
32. Grant MA, Bixley PF (2011) Geothermal reservoir engineering. Elsevier Science Ltd, Oxford.
33. Guéguen Y, Palciauskas V (1994) Introduction to the physics of rocks. Princeton University Press, Princeton.
34. Guéguen Y, Schubnel A (2003) Elastic wave velocities and permeability of cracked rocks. Tectonophysics 370:163-176
35. Gupta H, Sukanta R (2006) Geothermal energy: an alternative resource for the 21st century. Elsevier B.V., Oxford.
36. Hardy H (1981) Applications of acoustic emission techniques to rock and rock structures: a state of the art review. In: Drnevich V, Gray R (eds) Acoustic emission in geotechnical engineering practice, American Society for Testing and Materials, University of Michigan, Ann Arbor.
37. Heap MJ, Faulkner DR (2008) Quantifying the evolution of static elastic properties as crystalline rock approaches failure. Int J Rock Mech Min Sci 45:564-573
38. Heap MJ, Vinciguerra S, Meredith PG (2009) The evolution of elastic moduli with increasing crack damage during cyclic stressing of a basalt from Mt. Etna volcano. Tectonophysics 471:153-160
39. Heap MJ, Lavallée Y, Laumann A, Hess K-U, Meredith PG, Dingwell DB, Huismann S, Weise F (2013) The influence of thermal-stressing (up to 1000°C) on the physical, mechanical, and chemical properties of siliceous-aggregate, high-strength concrete. Construct Build Mater 42:248-265

40. Heap MJ, LavalleeY, Petrakova L, Baud P, ReushcleT, Varley NR, Dingwell DB (2014) Microstructural controls on the physical and mechanical properties of edifice-forming andesites at Volcán de Colima Mexico. J Geophys Res 119:2925-2963

41. Heard HC, Page L (1982) Elastic moduli, thermal expansion, and inferred permeability of two granites to 350°C and 55 megapascals. J Geophys Res 87:9340-9348

42. Hedenquist JW, Mroczek EK, Giggenbach WF (1988) Geochemistry of the Rotokawa geothermal system: summary of data, interpretation and appraisal for energy development. In: Chemistry Division DSIR Technical Note 88/6. , .

43. Hole HM (2013) Geothermal drilling - keep it simple. In: Proceedings of the 35th New Zealand geothermal workshop. , Rotorua, New Zealand.17–20 November 2013

44. Horie T, Muto T (2010) The world's largest single cylinder geothermal power generation unit - Nga Awa Purua Geothermal Power Station, New Zealand. Geothermal Res Council Trans 34:1039-1044

45. Jafari A, Babadagli T (2011) Effective fracture network permeability of geothermal reservoirs. Geothermics 40:25-38

46. Jaya MS, Shapiro SA, Kristinsdóttir LH, Bruhn D, Milsch H, Spangenberg E (2010) Temperature dependence of seismic properties in geothermal rocks at reservoir conditions. Geothermics 39:115-123

47. Ju Y, Yang Y, Peng R, Mao L (2013) Effects of pore structures on static mechanical properties of sandstone. J Geotech Geoenvironmental Eng 139:1745-1755

48. Kahraman S (2001) Evaluation of simple methods for assessing the uniaxial compressive strength of rock. Int J Rock Mech Min Sci 38:981-994

49. Kahraman S, Gunaydin O, Fener M (2005) The effect of porosity on the relation between uniaxial compressive strength and point load index. Int J Rock Mech Min Sci 42:584-589

50. Keshavarz M, Pellet FL, Loret B (2010) Damage and changes in mechanical properties of a gabbro thermally loaded up to 1,000°C. Pure Appl Geophys 167:1511-1523
51. Klinkenberg LJ (1941) The permeability of porous media to liquids and gases. Drilling and production practice. American Petroleum Institute, New York.
52. Kristinsdóttir LH, Flóvenz ÓG, Árnason K, Bruhn D, Milsch H, Spangenberg E, Kulenkampff J (2010) Electrical conductivity and P-wave velocity in rock samples from high-temperature Icelandic geothermal fields. Geothermics 39:94-105
53. Krupp RE, Seward TM (1987) The Rotokawa geothermal system, New Zealand; an active epithermal gold-depositing environment. Econ Geol 82:1109-1129
54. LavalléeY, Benson PM, Heap MJ, Hess KU, Flaws A, Schillinger B, Meredith PG, Dingwell DB (2013) Reconstructing magma failure and the degassing network of dome-building eruptions. Geology 41:515-518
55. Legmann H, Sullivan P (2003) The 30 MW Rotokawa I geothermal project five years of operation. In: International geothermal conference. , Reykjavik, Iceland.September 2003
56. Li L, Aubertin M (2003) A general relationship between porosity and uniaxial strength of engineering materials. Can J Civ Eng 30:644-658
57. Lim SS, Martin CD, Åkesson U (2012) In-situ stress and microcracking in granite cores with depth. Eng Geol 147–148:1-13
58. Lion M, Skoczylas F, Ledésert B (2005) Effects of heating on the hydraulic and poroelastic properties of Bourgogne limestone. Int J Rock Mech Min Sci 42:508-520
59. Luping T (1986) A study of the quantitative relationship between strength and pore-size distribution of porous materials. Cem Concr Res 16:87-96
60. Lutz SJ, Hickman S, Davatzes N, Zemach E, Drakos P, Robertson-Tait A (2010) Rock mechanical testing and

petrologic analysis in support of well stimulation activities at the Desert Peak Geothermal Field, Nevada. In: Proceedings of the thirty-fifth workshop on geothermal reservoir engineering. Stanford University, Stanford, California.1–3 February 2010

61. Martin CD (1993) The strength of Massive Lac du Bonnet granite around underground openings. Dissertation, University of Manitoba.
62. Martin CD, Chandler NA (1994) The progressive fracture of Lac du Bonnet granite. Int J Rock Mech Min Sci 31:643-659
63. Martínez-Martínez J, Benavente D, García-del-Cura MA (2011) Spatial attenuation: the most sensitive ultrasonic parameter for detecting petrographic features and decay processes in carbonate rocks. Eng Geol 119:84-95
64. Massiot C, McNamara D, Lewis B, Price L, Bignall G (2012) Statistical corrections of fracture sampling bias in boreholes from acoustic televiewer logs. In: New Zealand geothermal workshop proceedings. , Auckland, New Zealand.19–21 November 2012
65. Mueller S, Melnik O, Spieler O (2005) Permeability and degassing of dome lavas undergoing rapid decompression: an experimental determination. Bull Volcanol 67:526-538
66. Nara Y, Meredith PG, Yoneda T, Kaneko K (2011) Influence of macro-fractures and micro-fractures on permeability and elastic wave velocities in basalt at elevated pressure. Tectonophysics 503:52-59
67. Nara Y, Morimoto K, Hiroyoshi N, Yoneda T, Kaneko K, Benson PM (2012) Influence of relative humidity on fracture toughness of rock: implications for subcritical crack growth. Int J Solids Struct 49:2471-2481
68. Nicksiar M, Martin CD (2012) Evaluation of methods for determining crack initiation in compression tests on low-porosity rocks. Rock Mech Rock Eng 45:607-617
69. Obara Y, Sakaguchi K, Nakayama T, Sugawara K (1992) Anisotropy effect on fracture toughness of rocks. Int J Rock Mech Min Sci Geomech 30:137 Abstract

70. Ouchterlony F (1990) Fracture toughness testing of rock with core based specimens. Eng Fract Mech 35:351-366
71. Palchik V (2013) Is there a link between the type of the volumetric strain curve and elastic constants, porosity, stress and strain characteristics? Rock Mech Rock Eng 46:315-326
72. Palchik V, Hatzor YH (2002) Crack damage stress as a composite function of porosity and elastic matrix stiffness in dolomites and limestones. Eng Geol 63:233-245
73. Pereira J-M, Arson C (2013) Retention and permeability properties of damaged porous rocks. Comput Geotech 48:272-282
74. Pola A, Crosta G, Fusi N, Barberini V, Norini G (2012) Influence of alteration on physical properties of volcanic rocks. Tectonophysics 566–567:67-86
75. Pola A, Crosta GB, Fusi N, Castellanza R (2014) General characterization of the mechanical behaviour of different volcanic rocks with respect to alteration. Eng Geol 169:1-13
76. Powell T (2011) Natural subsidence at the Rotokawa Geothermal Field and implications for permeability development. Geothermal Res Council Trans 35:973-976
77. Quinao J, Sirad-Azwar L, Clearwater J, Hoepfinger V, Le Brun M, Bardsley C (2013) Analyses and modeling of reservoir pressure changes to interpret the Rotokawa Geothermal Field response to Nga Awa Purua Power Station operation. In: Proceedings of the 38th workshop on geothermal reservoir engineering. Stanford University, Stanford, California.11–13 February 2013
78. Rae A (2007) Rotokawa geology and geophysics. GNS Science consultancy report 2007/83 May 2007. GNS Science, Lower Hutt.
79. Rae AJ, McCoy-West AJ, Ramirez LE, Alcaraz SA (2009) Geology of production well RK28. Rotokawa Geothermal Field. GNS Science consultancy report 2009/253 September 2009. GNS Science, Lower Hutt.

80. Rae AJ, McCoy-West AJ, Ramirez LE, McNamara D (2010) Geology of production wells RK30L1 and RK30L2, Rotokawa Geothermal Field. GNS Science consultancy report 2010/02 January 2010. GNS Science, Lower Hutt.
81. Rajabzadeh MA, Moosavinasab Z, Rakhshandehroo G (2011) Effects of rock classes and porosity on the relation between uniaxial compressive strength and some rock properties for carbonate rocks. Rock Mech Rock Eng 45:113-122
82. Ramirez LE, Hitchcock D (2010) Geology of production well RK27L2, Rotokawa Geothermal Field, GNS Science Consultancy Report 2010/100 April 2010. GNS Science, Lower Hutt.
83. Reuschlé T, Gbaguidi Haore S, Darot M (2006) The effect of heating on the microstructural evolution of La Peyratte granite deduced from acoustic velocity measurements. Earth Planet Sci Lett 243:692-700
84. Richter D, Simmons G (1977) Microcracks in crustal igneous rocks: microscopy. In: Heacock JG, Keller GV, Oliver JE, Simmons G (eds) The earth's crust, American Geophysical Union, Washington, DC.
85. Rowland JV, Sibson RH (2004) Structural controls on hydrothermal flow in a segmented rift system, Taupo Volcanic Zone, New Zealand. Geofluids 4:259-283
86. Rowland JV, Wilson CJN, Gravley DM (2010) Spatial and temporal variations in magma-assisted rifting, Taupo Volcanic Zone, New Zealand. J Volcanol Geotherm Res 190:89-108
87. Rust AC, Manga M, Cashman KV (2003) Determining flow type, shear rate and shear stress in magmas from bubble shapes and orientations. J Volcanol Geotherm Res 122:111-132
88. Rutter EH (1986) On the nomenclature of mode of failure transitions in rocks. Tectonophysics 122:381-387
89. Saar MO, Manga M (1999) Permeability-porosity relationship in vesicular basalts. Geophys Res Lett 26:111-114

90. Schöpfer MPJ, Abe S, Childs C, Walsh JJ (2009) The impact of porosity and crack density on the elasticity, strength and friction of cohesive granular materials: insights from DEM modelling. Int J Rock Mech Min Sci 46:250-261
91. Sewell SM, Cumming WB, Azwar L, Bardsley C (2012) Integrated MT and natural state temperature interpretation for a conceptual model supporting reservoir numerical modelling and well targeting at the Rotokawa Geothermal Field, New Zealand. In: Proceedings of the thirty-seventh workshop on geothermal reservoir engineering. Stanford University, Stanford California.30 January to 1 February 2012
92. Sewell SM, Cumming W, Bardsley CJ, Winick J, Quinao J, Wallis IC, Sherburn S, Bourguignon S, Bannister S (2013) Interpretation of microearthquakes at the Rotokawa Geothermal Field, 2008 to 2012. In: Proceedings of the 35th New Zealand geothermal workshop. , Rotorua, New Zealand.17–20 November 2013
93. Sherburn S, Bourguignon S, Bannister S, Sewell S, Cumming B, Bardsley C, Quinao J, Wallis I (2013) Microseismicity at Rotokawa Geothermal Field, 2008 to 2012. In: Proceedings of the 35th New Zealand geothermal workshop. , Rotorua, New Zealand.17–20 November 2013
94. Siega C, Grant M, Powell T (2009) Enhancing injection well performance by cold water stimulation in Rotokawa and Kawerau geothermal field. In: Proceedings of PNOC-EDC conference. , Manila, Philippines.27–28 September 2009
95. Smith R, Sammonds PR, Kilburn CRJ (2009) Fracturing of volcanic systems: experimental insights into pre-eruptive conditions. Earth Planet Sci Lett 280:211-219
96. Sousa LMO, Suárez del Río LM, Calleja L, Ruiz de Argandoña VG, Rey AR (2005) Influence of microfractures and porosity on the physico-mechanical properties and weathering of ornamental granites. Eng Geol 77:153-168
97. Stanchits S, Vinciguerra S, Dresen G (2006) Ultrasonic velocities, acoustic emission characteristics and crack damage of basalt and granite. Pure Appl Geophys 163:975-994

98. Stimac JA, Powell TS, Golla GU (2004) Porosity and permeability of the Tiwi geothermal field, Philippines, based on continuous and spot core measurements. Geothermics 33:87-107

99. Stimac J, Nordquist G, Suminar A, Sirad-Azwar L (2008) An overview of the Awibengkok geothermal system, Indonesia. Geothermics 37:300-331

100. Takarli M, Prince W, Siddique R (2008) Damage in granite under heating/cooling cycles and water freeze thaw condition. Int J Rock Mech Min Sci 45:1164-1175

101. Tugrul A, Gurpinar O (1997) The effect of chemical weathering on the engineering properties of Eocene basalts in northeastern Turkey. Environ Eng Geosci 3:225-234

102. Tutluoglu L, Keles C (2011) Mode I fracture toughness determination with straight notched disk bending method. Int J Rock Mech Min Sci 48:1248-1261

103. Ulusay R, Hudson J (2007) The complete ISRM suggested methods for rock characterization, testing and monitoring: 1974–2006. Elsevier, Antalya, Turkey.

104. Underwood EE (1970) Quantitative stereology for microstructural analysis. In: Underwood EE (ed) Quantitative stereology, Addison-Wesley, Reading, Massachusetts.

105. Vernik L, Bruno M, Bovberg C (1993) Empirical relations between compressive strength and porosity of siliciclastic rocks. Int J Rock Mech Min Sci Geomech 30:677-680 Abstract

106. Vinciguerra S, Trovato C, Meredith P, Benson P (2005) Relating seismic velocities, thermal cracking and permeability in Mt. Etna and Iceland basalts. Int J Rock Mech Min Sci 42:900-910

107. Wallis I, McCormick S, Sewell S, Boseley C (2009) Formation assessment in geothermal using wireline tools - application and early results from the Ngatamariki Geothermal Field, New Zealand. In: Proceedings of the New Zealand Geothermal Workshop. , Rotorua, New Zealand.16–18 November 2009

108. Walsh JB (1965) The effect of cracks on the compressibility of rock. J Geophys Res 70:381-389
109. Walsh JB (1965) The effect of cracks in rocks on Poisson's ratio. J Geophys Res 70:5249-5257
110. Wang HF, Bonner BP, Carlson SR, Kowallis BJ, Heard HC (1989) Thermal stress cracking in granite. J Geophys Res 94:1745-1758
111. Watanabe T, Shimizu Y, Noguchi S, Nakada S (2008) Permeability measurements on rock samples from Unzen Scientific Drilling Project Drill Hole 4 (USDP-4). J Volcanol Geotherm Res 175:82-90
112. Wilson CJN, Houghton BF, Mcwilliams MO, Lanphere MA, Weaver SD, Briggs RM (1995) Volcanic and structural evolution of Taupo Volcanic Zone, New Zealand: a review. J Volcanol Geotherm Res 68:1-28
113. Wong TF, Baud P (2012) The brittle-ductile transition in porous rock: a review. J Struct Geol 44:25-53
114. Wu XY, Baud P, Wong TF (2000) Micromechanics of compressive failure and spatial evolution of anisotropic damage in Darley Dale sandstone. Int J Rock Mech Min Sci 37:143-160

Chapter 5

Electromagnetic Heating of Heavy Oil and Bitumen: A Review of Experimental Studies and Field Applications

Albina Mukhametshina[1, 2] and Elena Martynova[1, 2]

[1]Harold Vance Department of Petroleum Engineering, Texas A&M University, 3116 TAMU-407 Richardson Building, College Station, TX, USA

[2]Gubkin Russian State University of Oil and Gas, 65 Leninsky Prospekt, Moscow, Russia

ABSTRACT

Viscosity is a major obstacle in the recovery of low API gravity oil resources from heavy oil and bitumen reservoirs. While thermal recovery is usually considered the most effective method for lowering viscosity, for some reservoirs introducing heat with commonly implemented thermal methods is not recommended. For these types of reservoirs, electromagnetic heating is the recommended solution. Electromagnetic heating targets part of the reservoir instead of heating the bulk of the reservoir, which means that the targeted area can be heated up more effectively and with lower heat losses than with other thermal methods. Electromagnetic heating is still relatively new and is not widely used as an alternate or addition to traditional thermal recovery methods. However, studies are being conducted and new technologies proposed that could help increase its use. Therefore, the objective of this study is to investigate the recovery of heavy oil and bitumen reservoirs by electromagnetic heating through the review of existing laboratory studies and field trials.

INTRODUCTION

High-frequency electromagnetic radiation is a relatively new technique for use in enhanced oil recovery methods. It has been tested by theoretic, laboratories and field trial research in Russia [1–10], the United States [11–17], Canada [18–21], and other countries [22–34]. Traditional thermal recovery and well stimulation techniques using hot steam or fluid are not effective in some cases [7, 35] due to prohibitive heat losses from injection wells and reservoirs, low reservoir injectivity (especially for bitumen deposits), steam leakage, large overburden heat loss at thin pay zones, permafrost conditions, and so forth. Furthermore, commonly used thermal recovery methods are not considered environmentally friendly, damaging the hydrogeologic environment and contributing to the greenhouse effect.

The most important thing in electromagnetic heating is that the heat is developed within the material rather than being brought from outside, which means the material is heated more uniformly throughout the medium [27]. Therefore, instead of heating the bulk reservoir volume, part of the reservoir can be targeted and heated more effectively with lower heat loss than other thermal methods. Unlike traditional thermal recovery methods, microwave heating causes friction by vibration of molecules, which results in dielectric heating of the reservoir. Heat and mass transfer in different environments under microwave influence was studied by a number of scientists around the globe, but its application as an EOR method is not yet fully understood. Microwave heating is not used productively because of the lack of reliable information about the processes of heat and mass transfer in a multiphase system in porous media under the influence of electromagnetic radiation, which does not allow effective control. Therefore, current research studies use modeling to discover optimum design parameters for the use of microwave heating in field applications.

REVIEW OF EXPERIMENTAL ELECTROMAGNETIC HEAT STUDIES

The success of near wellbore heating with electromagnetic waves has been proven experimentally [1]. To represent reservoir rock, quartz sand with a 7.7 dielectric constant and a 0.083 tangent was used in the laboratory. A 20% initial water saturation and an 80% initial oil saturation with 16.61 cps (16.61×10^{-3} Pa·s) gravity and 0.86 g/cm^3 (860 kg/m^3) density oil were maintained. The dielectric constant and the loss tangent of the oil sample were 2.23 and 0.019, respectively. In this study of electric and high-frequency electromagnetic heating of a reservoir model, the temperature of the medium was measured by thermometers located at different points of the experimental setup. In another case, a linear radiator with a length of 87 cm and a diameter of 19 mm was placed in the center of the setup. The linear radiator was connected via acoaxial cable to the generator supplying electromagnetic waves with a frequency of 13.56 MHz. In the experiment, it was discovered that when fluid temperature was exposed to a high-frequency electromagnetic field at the same distance from the radiation source it was greater than one initiated by electric heating. In this case, the thermal conductivity of the medium is affected only slightly. When heat is induced throughout the whole volume of the medium, the amount of heat introduced depends largely on the electrical properties of the medium. The study summarized the advantages of microwave heating over electrical heating as having deeper penetration, quicker heating, and lower heat losses.

Sayakhov [2] discussed the physical foundations of fluid filtration in high-frequency fields. Radial fluid filtration through porous media in a heterogeneous high-frequency electromagnetic field was studied experimentally. A specially designed core holder, filled with a representative porous medium, was located in a coaxial resonator. A high-frequency wave generator was used to generate mw energy, which was directed to the resonator through cables. Studied liquid came into the porous media through the hose first, then through the lumen of the inner conductor of the resonator dripped into a graduated cylinder. Experiments were started with a 200–300 watts power microwave oven at 2400 MHz frequency and 500 watts vibrational power. Temperatures were recorded throughout the experiments via thermocouples inserted in the core holder. Kerosene was used as a representative reservoir fluid. The measurements were performed with and without the influence of a high-frequency electromagnetic field. It was established that exposure to high-frequency electromagnetic fields leads to a sharp increase in flow rate per unit of time and fluid temperature at the outlet. In addition, the flow rate increases dramatically once exposed to the electromagnetic field, while the temperature increases after 10 seconds. After the field is discontinued, a sharp decrease in the flow rate is observed and a gradual cooling of the porous medium takes place.

Experimental studies [3] on the influence of electromagnetic fields with a frequency of $3*10^5$Hz to $6*10^5$ Hz on thermal conductivity of dielectric liquids showed that the thermal conductivity of the liquids increased when exposed. The thermal conductivity increases as the magnitude of the dipole moment of the liquid used in the experiment parameters increases with the frequency and intensity of field.

Fatikhov [5] conducted experimental research on the flow of bitumen oil at different pressure gradients in a high-frequency electromagnetic field. The experiment focused on changes in the volumetric flow rate of filtered oil under different pressures at different temperatures. The initial pressure drop for bitumen from the Mordovo-Karmalskoye deposit in reservoir conditions was 0.003 MPa/m. During the experiment, the pressure drop decreased rapidly as the temperatures were increased, resulting in bituminous oil becoming a Newtonian liquid. Therefore, it was established that the application of electromagnetic heating improves fluid flow behavior and the non-Newtonian properties of bitumen decrease rapidly with increasing temperature.

In 1992, Kasevich et al. [17] studied electromagnetic heating of rock samples at 1 kW (frequency 50.55 MHz) and 200 W (144 MHz) and found that a particular type of rock could be heated to 423 K when exposed to an RF electromagnetic field. The rock, which had low thermal conductivity, heats poorly when hot steam is pumped into the reservoir. They conducted experiments at both normal and formation pressures.

Ovalles et al. [30] used a microwave with 650 watts power to heat core samples saturated with oil of 25 API gravity and 7.7 API gravity (a sample from the Orinoco River Basin). Medium API oil temperatures were measured in 0.5, 1, and 1.5 minute intervals and heavy oil intervals were increased to 1, 5, and 10 minutes. The experimental results were used to test mathematical models and predict the production of three abstract oil reservoirs in Venezuela.

Chakma and Jha [23] conducted laboratory experiments using electromagnetic heating on a scaled thin heavy oil reservoir pay zone model. Gas injection with horizontal wells during electromagnetic heating was achieved. The aim was to decrease oil viscosity with electromagnetic heating and obtain a gas drive with the injected gas. Using nitrogen for the injected gas, they were able to prove that for thin pay zones heating of the wellbore vicinity is sufficient, by achieving oil recoveries as high as 45% of original oil in place compared to estimated primary recovery rates of less than 5%. Recovery achieved by use of the combined method was higher than that of nitrogen injection or electromagnetic heating alone. Chakma and Jha also discussed a number of parameters affecting the results of the combined method, including

- gas injection pressure (when no gas was injected, oil was produced only due to gravity drainage and no significant convective transport occurred; therefore, gas injection provided an oil rate increase with the increasing injection pressure);
- temperature (the initial production rate was not significantly affected by temperature, but later there was an increase in the production rate, meaning that overall recovery increased with temperature);
- electromagnetic frequency (the higher the frequency, the greater the recovery);
- oil viscosity (as expected, a higher oil viscosity leads to a lower recovery for a given electromagnetic frequency, temperature, and gas injection pressure);
- salinity (higher salinity provides higher recovery due to the higher conductivity of saline water compared with distilled water);
- electrode distance (recovery is similar, but closer electrode spacing provides faster production rates) [23].

Hascakir et al. [24] conducted a laboratory study of microwave-assisted gravity drainage on heavy oil samples from reservoirs in Turkey (Bati Raman, 9.5 API; Garzan, 12 API; and Camurlu, 18 API) using a specially designed novel graphite core holder packed with crushed limestone. Their study described the effects of operational parameters like heating time, waiting period and rock, and fluid properties on the effectiveness of microwave heating. Some of the conclusions made are confirming ones found in Chakma and Jha [23], like the positive effect of high water salinity and water saturation. Hascakir et al. [24] also concluded that water wet conditions are preferable for obtaining higher oil recoveries and that large porosity and permeability are also favorable. When microwave heating is applied to oil samples continuously, higher temperatures are reached, which allows better results to be achieved than with periodic heating when microwave heating is applied for a limited time in periodic intervals. Therefore, higher temperatures allow for better results in continuous heating.

Jha et al. [27] proposed using microwave-assisted gravity drainage (MWAGD) in the Mehsana oil field in India. They heated specially prepared samples with the required characteristics from that field in the laboratory using a microwave with variable power up to 1000 watts operating at 3 GHz frequency, which allowed them to obtain temperature and viscosity profiles of the gravity-drained oil. They described effects of initial oil and water saturations, wettability, porosity, and permeability similar to those found by Hascakir et al. [24] and Chakma and Jha [23].

Jha et al. [27] suggested using MWAGD commercially by drilling one horizontal well and multiple vertical ones with downhole microwave antennas. However, this might not allow deep enough heat penetration, so other options are also proposed such as a combination of two horizontal wells and installing antenna inside the horizontal production well. Vertical separation of the horizontal well pair is approximately 15 meters, which is far more efficient than SAGD in which the separation is around 5 meters. Because crude oil absorbs microwave heat weakly, Jha et al. also proposed increasing thermal conductivity by injecting powdered metallic oxides, chlorides, or activated carbon through a fracture operation. The working principle and description of laboratory applications of such additives to heavy oil can be found in various studies by Hascakir et al., Kershaw et al., and Odenbach [25, 28, 29]. Technical principles of the SAGD method assisted by electromagnetic heating (EM-SAGD process) were reported by Koolman et al. [26]. Inductive heating was initiated in the laboratory using an EM source with a working frequency of 142 kHz. The sample was heated for 10 minutes at a power of 7.2 kW, achieving a rise in the temperature of 7.5 K. Laboratory and field processes were modeled using a numerical simulator, combining electromagnetic and thermal modules. It was specially built and can be applied to field-scale simulations. According to simulation results, a 38% increase in bitumen production was predicted compared to conventional SAGD.

Kovaleva et al. investigated the effects of radio frequency electromagnetic (RF-EM) fields and electrical heating on the mass- and heat-transfer processes in a multicomponent hydrocarbon system flowing in porous media [4, 6]. Three different types of experiments were carried out: solvent (kerosene) flooding under the RF-EM field, solvent (kerosene) flooding under electrical heating, and cold solvent (kerosene) flooding. In all three experiments, the physical characteristics of the model and heating conditions (temperatures) were identically maintained. Two series of experiments on models with different granulometric composition of formation were also carried out.

Figure 1 shows the dependence of oil recovery on the volume of solvent injected. This figure demonstrates that the highest oil recovery was obtained by applying an RF-EM field. Kovaleva et al. [4, 6] concluded that following RF-EM influence on the oil-saturated samples the quantity of the received oil is more than the quantity received under electrical (and thermal) processing at identical temperatures of heating of the media. It confirms the additional "nonthermal" action of the electromagnetic field.

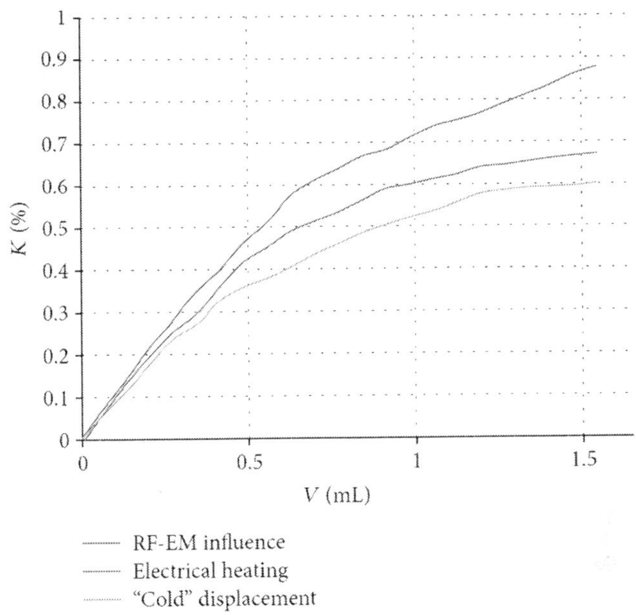

Figure 1: Dependence of oil recovery (K) on the relative volume of solvent injected. Adapted from—[6].

TECHNOLOGIES OF IN SITU ELECTROMAGNETIC HEATING OF HEAVY OIL AND BITUMEN

The first production method applying microwave heating to well production was patented in 1956 [14]. Electromagnetic waves were transferred to the well bottom from the surface through a coaxial system of internal and external pipes (tubing and casing). Interaction of electromagnetic waves with the formation causes the emergence of distributed volumetric heat sources and reduces the viscosity of the reservoir fluid. In 1965, Haagensen [13] described a device for generating high-frequency electromagnetic waves at the mouth of the well and a method of delivering electromagnetic energy through coaxial lines and waveguides to the bottom hole. In 1987, Wilson [16] described a similar device in his work, with some modifications of the radiating element of EM waves.

A huge drawback to the methods described by Haagensen and Wilson [13, 16] is the shallow penetration of electromagnetic waves, and, hence, a low sweep efficiency of heating. When the method described by Ritchey [14] was implemented, there were large losses of electromagnetic energy. Due to the finite conductivity of tubing, they are heated and electromagnetic energy is dissipated in rocks surrounding the well, resulting in large wellbore heat losses, especially if a permafrost layer is present.

Sayakhov et al. [8] proposed a method of recovery that included creating a combustion front with a simultaneous electromagnetic current influence. The environment is heated when exposed to an electromagnetic field, which decreases the viscosity and increases the mobility of crude oil. It is assumed in this method that the EM field continues influencing the reservoir after the combustion is initiated.

Review of Electromagnetic Heating Field Studies

Electromagnetic heating field trials have been carried out in Russia (Bashkortostan and Tatarstan) [2, 9, 10], the United States (California and Utah) [11, 17], and in Canada (Alberta and Saskatchewan) [18–21].

Russia

In Russia, field tests of radio frequency electromagnetic heating of the near-wellbore zone were first launched in 1969 at Well 40/19 in the Ishimbayskoye Oil Field in Bashkortostan and continued in the Yultimirovskoye Bitumen Field in Tatarstan according to Sayakhov et al. [2, 9, 10]. Characteristics of Well 40/19 from the Ishimbayskoye Oil Field are represented in Table 1.

Table 1: Characteristics of Well 40/19 of the Ishimbayskoye Oil Field

Parameter	Value
Depth, m	830
Casing diameter, in.	6
Tubing diameter, in.	2
Flow rate, ton/day	3
Well temperature, K	287–289
Paraffin content, %	2.3
Resins content, %	11
Density, kg/m^3	890
Viscosity at 20°C, m^2/c	20 * 10^{-6}

Sayakhov et al. [10].

The source of high-frequency electromagnetic energy was a generator providing an optimum oscillation output power of 63 kW at a frequency of 13.56 MHz. Standard RF coaxial cable was used to supply high-frequency electromagnetic energy from the generator to the well. Temperature measurements were carried out using a thermograph at the bottom hole. Temperature was recorded continuously at a fixed depth of 650–655 m (in the open hole, where the radiating element was working) while the generator was operating (Table 2).

Table 2: Dynamics of temperature growth

Heating time, days	Temperature increase at the bottom hole, K
0.5	283
1	290
2	300
3	306
4	311
5	313

Sayakhov et al. [9].

Yultimirovskoye Bitumen Field

RF electromagnetic heating was conducted in the Yultimirovskoye Bitumen Field by Sayakhov in 1980 [2] (Table 3). Two wells spaced 5 meters apart were studied, Well 150 and Well 1.

Table 3: Characteristics of the Yultimirovskoye reservoir

Parameter	Value
Porosity, %	25
Bitumen saturation, %	3.6
Permeability, micm^2	0–0.183

Sayakhov [2].

Electromagnetic heating of the bitumen reservoir was conducted in several stages at different conditions. Initially, the RF-EM installation was set to about of 20 kW power. After 36.5 hours, the temperature at the bottom of Well 150 has increased from 282 to 389 K. No temperature change was observed in Well 1. In the next phase, RF-EM installation was reset to approximately 30 kW of power. As a result, the temperature in Well 150 reached 423 K after six hours. It should be noted that the growth rate of temperature increased. In the third phase, the RF-EM installation was set to a maximum of 60 kW, which caused the heating intensity in Well 150 to increase greatly. After 5.5 hours, the temperature in Well 150 increased from 417 to 463 K. Next, the RF-EM unit was turned off for 2 hours, resulting in a temperature drop to 423 K.

During the next 32 hours, the RF-EM installation operated at maximum output, causing the heating of the bottom hole of Well 150 up to 583 K and the bottom hole of Well 1 up to 318 K. In Sayakhov's experiment [2], the fluoroplastic collars that centered the tubing in Well 150 melted (their maximum operating temperature was 573 K). As a result, a short circuit between the casing and tubing occurred and the RF-EM unit broke and was disabled. Before this disruption, the RF-EM installation worked steadily throughout the field experiment.

Deep heat penetration (up to 5 m in the reservoir) was demonstrated by temperature measurements done during the cooling of Well 150's drain zone after the exposure. The temperature decreased from 373 K to 343 K in three days. After a few days of electromagnetic heating, wellbore heat distribution revealed low heat losses to both the overburden and underburden.

The United States

Bakersfield, California, United States

In 1992, Kasevich et al. [17] conducted field tests of RF-EM heating in the United States in the Bakersfield, California, field. The goal was to prove the concept that controlled RF-EM radiation could be used as a thermal EOR method. The production of reservoir fluids was not measured because it was a quality, not quantity, study designed to gain a better understanding of underground processes.

A high-frequency electromagnetic wave generator with a capacity of 25 kW and frequency of 13.56 MHz was used to heat the reservoir at Well 100D. Heat penetration was determined by temperature measurements in the surrounding observation wells T10, T20, and T30, which were located 3, 6, and 9 meters from Well 100D. Kasevich et al. [17] also proved that the RF producer used could efficiently focus its radiation pattern into the desired region by measuring return loss and electromagnetic radiation. In well T10, situated 3 meters from Well 100D, the medium temperature was increased from 293 K to 393 K over 20 hours of EM heating.

Avintaquin Canyon and Asphalt Ridge (Utah)

In 1980, one of the most detailed studies on electromagnetic-heat-based oil recovery was done at the Illinois Institute of Technology

Research Institute (IITRI) by Bridges et al. [11]. They carried out extensive research work on the use of the different types of electromagnetic heating for different types of deposits, oil shale, and tar sand.

Bridges et al. [11] tested their IITRI technique of RF electromagnetic heating with two field experiments in Avintaquin Canyon, Utah, USA. Shale that was 6 meters thick was found in outcrops convenient for relatively cheap horizontal experiments. So arrays of holes were drilled and electrodes were inserted to a depth of 1 meter. These tests allowed the researchers to gain experience and to prove it was possible to achieve in situ pyrolysis of oil shale, thus increasing its thermal maturity. The power applied to the formation ranged from 5 kW to 20 kW, with a frequency of 13.56 MHz. As a result of EM heating, temperatures rose to 673 K and 20–30 % of the oil content was collected. However, it should be noted that the amount of produced oil was badly affected by the presence of cracks which allowed light hydrocarbons to escape (evaporate).

In 1981, Bridges et al. [11] conducted field tests on tar sand at Asphalt Ridge, Utah, USA. The first experiment tested the gravity drive bitumen recovery process, designed to prove EM heating concepts and improve equipment design. This experiment used vertical electrode placement and a mined collection chamber and tunnel. It was equipped with a 200 kW radio transmitter and heated 25 m^3 of tar sand. In the first experiment, the roof of the mined chamber was not supported well enough, which resulted in early termination of the experiment. Therefore, the situation had to be fixed by constructing a concrete arch for the second pilot test. The heating power used on tar sand varied from 40 kW to 75 kW with a frequency of 13.56 MHz.

The second test quantified the results of heating over a longer period and at higher temperatures. In this experiment, temperatures exceeded 473 K, and 30 to 35% recovery was achieved in just 20 days. This was encouraging because continuation of heating could have resulted in even higher recovery. The power loss was minimal in all the experiments, which proved the efficiency of heating.

CANADA

The Wildmere Field, Alberta, Canada

According to Spencer [18] commercial EM heating was first introduced in the field at Wildmere, Alberta, Canada (Table 4). The first well was drilled in January 1986 and began producing oil in March of the same year. Before EM heating began in May, the well was producing about 0.95 tonnes/day. After EM heating commenced, production rates increased and soon settled at the level of 3.18 tonnes/day until November 1986, when the well was closed due to technical reasons. Another well in this field increased in production from 1.59 tonnes/day to an average of 4.77 tonnes/day, with the maximum flow rate reaching 9.54 tonnes/day.

Table 4: Characteristics of the reservoir in the Wildmere field, Alberta, Canada

Parameter	Value
Net thickness, m	1
Depth, m	600
Density, kg/m^3	987
Oil viscosity at 20°C, Pa·s	20

Spencer [18].

The Lloydminster Heavy Oil Area, Saskatchewan, Canada

In 1988-1989, two electromagnetic stimulation projects were conducted by Davidson [22] in the Lloydminster heavy oil area in Saskatchewan, Canada. Unfortunately the economic potential of the process could not be evaluated from either pilot test, because long-term heating could not be achieved due to equipment failure (casing insulation) and special reservoir conditions. However, the technical results looked promising.

The first pilot well was in Northminster (Saskatchewan, Canada) and produced 11.4 API oil from the Sparky formation. The power was applied to the well in a pulsating manner with a baseline of 20 kW with four-hour spikes of 30 kW (2 daily), in order to reduce the risk of significant damage to the insulation. Later, the power was increased to a 25 kW baseline with 35 kW pulses and finally to a 30 kW baseline with 50 kW peaks. At that point the insulation failed, and the power rates came down to 28 kW. Power rates were later leveled at 47 kW and stayed that way until terminated.

As can be seen from Table 5, water cut and production both reacted positively to electromagnetic stimulation. However, it should be noted that some portion of the increased production is related to the increase in pump speed. Water cut drop can be related directly to the EM effect and improvement in oil mobility. Once the heating has been terminated, technical parameters return rapidly to their initial states.

Table 5: Oil field performance for the Northminster pilot: primary and achieved by EM heating

Parameter	**Primary**	**EM heating**
Production rate, m^3/day	10–12	20
Water cut, %	15–20	10–12
Productivity index, bbl/psi	0.33	0.42
Stimulation ratio		1.27

Davidson [22].

The second pilot well was situated in Lashburn (Saskatchewan, Canada) and produced very viscous 11.4 API crude oil from the Sparky formation. During the reservoir heating phase, the power ranged from 13 to 18 kW. This well has high sand cuts and before electrical power was applied its production was not stable and regularly had to be stimulated by flushing the wellbore. The peak production reached 5.0 m^3/day and had begun to decline before electromagnetic heating was applied. Electromagnetic heating reduced the water cut, but the well was still prone to high water cuts after shut-in periods. Oil production also increased when electromagnetic forces were applied, until it reached 9 m^3/day. During the initial heating phase of the reservoir, the temperature in the bottom hole increased steadily from 295 K to 309 K, but temperatures began dropping immediately after the power was turned off or when power delivery systems failed.

ESEIEH (Alberta, Canada)

Enhanced Solvent Extraction Incorporating Electromagnetic Heating technology (ESEIEH) has been patented and is currently undergoing tests, as reported by Rassenfoss [20]. The ESEIEH consortium is relying on three oil company partners to help with this testing: Laricina Energy, Nexen, and Suncor Energy. The pilot project is planned to take three years and currently is in the first stage. Field application is expected to start later in 2013. The ESEIEH method combines the familiar horizontal well pair design commonly used in the Canadian oil sands, coupled with heating using RF-EM waves and solvents, such as butane or propane. The company aims to heat the reservoir by running an antenna underground that emits enough energy to raise the temperature to 50°C (120°F).

CONCLUSIONS

A review of electromagnetic heating for enhanced oil recovery was presented in this paper. A number of studies show that electromagnetic heating is a promising method of enhanced oil recovery. However, the studies to date are limited, and only a few field trials have been reported. Most of the current research is based on laboratory experiments or numerical models. It should be noted that this paper did not cover the computer simulations carried out to research the effectiveness of EM heating. Better understanding of the in situ electromagnetic process is essential and can be achieved by combining laboratory, numerical, and field-scale tests. At the moment it is not possible to assess the efficiency of EM heating or the opportunities for economic applications of it alone or in combination with traditional methods; therefore, more global studies should be conducted.

Even though sustainability of this technology has not yet been completely evaluated, the method definitely should not be overlooked by the industry because of its enormous potential. Attempts should be made to develop viable screening criteria for possible production of heavy oil, oil shale, and tar sand deposits.

REFERENCES

1. S. Chistyakov, F. Sayakhov, and G. Balabyan, "Experimental study of formations dielectric properties under the influence of high-frequency electromagnetic fields," in University Investigations: Geology and Exploration, pp. 153–156, 1971.
2. F. Sayakhov, "Particular properties of filtration and fluid flow under the influence of high-frequency electromagnetic field," in Joint University Scientific Book, pp. 108–120, 1980.
3. B. Savinikh, V. Dyakonov, and A. Usmanov, "The influence of alternating electric currents on the thermal conductivity of dielectric fluids," Journal of Engineering Physics and Thermophysics, no. 2, pp. 269–276, 1981 (Russian).
4. A. Davletbaev and L. Kovaleva, "Combined RF EM/solvent treatment technique: heavy/extra-heavy oil production model case study," in Proceedings of the 10th Annual International Conference Petroleum Phase Behavior and Fouling, Rio de Janeiro, Brazil, 2009.
5. M. A. Fatikhov, "Experimental study of bitumen initial pressure gradient in the electromagnetic field,"University Investigations: Oil and Gas, no. 5, pp. 93–94, 1990 (Russian).
6. L. Kovaleva, A. Davletbaev, T. Babadagli, and Z. Stepanova, "Effects of electrical and radio-frequency electromagnetic heating on the mass-transfer process during miscible injection

for heavy-oil recovery,"Energy and Fuels, vol. 25, no. 2, pp. 482–486, 2011. · ·

7. G. Malofeev, O. Mirsaetov, and I. Cholovskaya, "Injection of hot fluids for enhanced oil recovery and well stimulation," in Regular and Chaotic Dynamics, Institute of Computer Science, RussiaIgevsk, Russia, 2008.
8. F. Sayakhov, R. Bulgakov, V. Dyblenko, B. Deshura, and M. Bykov, "About HF heating of bitumen reservoirs," Petroleum Engineering, no. 1, pp. 5–8, 1980 (Russian).
9. F. L. Sayakhov, L. A. Kovaleva, M. A. Fatikhov, and G. A. Khalikov, "Method of thermal effect on oil-bearing formation," SU Patent 1723314, 1992.
10. F. Sayakhov, I. Habibullin, M. Yagudin, and M. Fatyhov, "Technique and technology of thermal well stimulation on the basis electro-thermo-chemical and electromagnetic effects," University Investigations: Oil and Gas, no. 2, pp. 33–42, 1992 (Russian).
11. J. E. Bridges, J. J. Krstansky, A. Taflove, and G. C. Sresty, "The IITRI in situ RF fuel recovery process,"Journal of Microwave Power, vol. 18, no. 1, pp. 3–14, 1983. ·
12. J. Bridges, "Method for in-situ heat processing of hydrocarbonaceous formation," US Patent 4140180, 1979.
13. A. D. Haagensen, "Oil well microwave tools," Patent USA 3170119, 1965.
14. H. W. Ritchey, "Radiation Heating System, US Patent," Tech. Rep. 2757738, 1956.
15. G. C. Sresty, R. H. Snow, and J. E. Bridges, "Recovery of liquid hydrocarbons from oil shale by electromagnetic heating in-situ," US Patent 4485869, 1984.
16. R. Wilson, "Well production method using microwave heating," US Patent 4485868, 1987.
17. R. S. Kasevich, S. L. Price, D. L. Faust, and M. F. Fontaine, "Pilot testing of a radio frequency heating system for enhanced oil recovery from diatomaceous earth," in Proceedings of the

SPE Annual Technical Conference & Exhibition, pp. 105–113, New Orleans, La, USA, September 1994.

18. H. L. Spencer, "Electromagnetic Oil Recovery, Ltd," Calgary, Canada, 1987.
19. F. E. Vermeulen and F. S. Chute, "Electromagnetic techniques in the in-situ recovery of heavy oils,"Journal of Microwave Power, vol. 18, no. 1, pp. 15–29, 1983. ·
20. S. Rassenfoss, "Seeking more oil, fewer emissions," Journal of Petroleum Technology, vol. 64, no. 9, pp. 34–38, 2012.
21. B. C. W. Mcgee and F. E. Vermeulen, "The mechanisms of electrical heating for the recovery of bitumen from oil sands," Journal of Canadian Petroleum Technology, vol. 46, no. 1, pp. 28–34, 2007. ·
22. R. J. Davidson, "Electromagnetic stimulation of Lloydminster heavy oil reservoirs: field test results,"Journal of Canadian Petroleum Technology, vol. 34, no. 4, pp. 15–24, 1995.
23. A. Chakma and K. N. Jha, "Heavy-oil recovery from thin pay zones by electromagnetic heating, paper SPE 24817," in Proceedings of the Annual Technical Conference and Exhibition, Society of Petroleum Engineers, Washington, DC, USA, October 1992. ·
24. B. Hascakir, C. Acar, Schlumberger, B. Demiral, and S. Akin, "Microwave assisted gravity drainage of heavy oils," in Proceedings of the International Petroleum Technology Conference (IPTC ‹08), pp. 1908–1916, Kuala Lumpur, Malaysia, December 2008.
25. B. Hascakir, T. Babadagli, and S. Akin, "Experimental and numerical modeling of heavy-oil recovery by electrical heating, paper SPE 117669," in Proceedings of the International Thermal Operations and Heavy Oil Symposium (ITOHOS ‹08), p. 14, Society of Petroleum Engineers, Alberta, Canada, October 2008. ·
26. M. Koolman, N. Huber, D. Diehl, and B. Wacker, "Electromagnetic heating method to improve steam assisted gravity drainage, paper 1177481," in Proceedings of the

International Thermal Operations and Heavy Oil Symposium (ITOHOS ‹08), pp. 327–338, Society of Petroleum Engineers, Alberta, Canada, October 2008.

27. K. A. Jha, N. Joshi, and A. Singh, "Applicability and assessment of micro-wave assisted gravity drainage (MWAGD) applications in Mehsana heavy oil field, paper SPE 14591," in Proceedings of the SPE Heavy Oil Conference and Exhibition, Society of Petroleum Engineers, Kuwait City, Kuwait, December 2011. ·

28. J. R. Kershaw, G. Barrass, and D. Gray, "Chemical nature of coal hydrogenation oils part I. The effect of catalysts," Fuel Processing Technology, vol. 3, no. 2, pp. 115–129, 1980. ·

29. S. Odenbach, "Ferrofluids—magnetically controlled suspensions," Colloids and Surfaces A, vol. 217, no. 1–3, pp. 171–178, 2003. · ·

30. C. Ovalles, A. Fonseca, A. Lara et al., "Opportunities of downhole dielectric heating in Venezuela: three case studies involving medium, heavy and extra-heavy crude oil reservoirs, paper SPE 78980," inProceedings of the International Thermal Operations and Heavy Oil Symposium and International Horizontal Well Technology Conference, Alberta, Canada, November 2002. ·

31. M. A. Ayrapetyan, "About oil fields development prospects by high-frequency currents electrical fields," in Materials of KSSR Institute of Oil, pp. 38–52, 1958.

32. M. A. Ayrapetyan, V. S. Velikanov, and E. Ya. Magnikov, "Reservoir high-frequency heating investigations," in Materials of KSSR Institute of Oil, pp. 113–124, 1959.

33. M. A. Carrizales, L. W. Lake, and R. T. Johns, "Production improvement of heavy-oil recovery by using electromagnetic heating, paper SPE 115723," in Proceedings of the SPE Annual Technical Conference and Exhibition (ATCE ‹08), Denver, Colo, USA, September 2008. · ·

34. A. D. Hiebert, F. E. Vermeulen, F. S. Chute, and C. E. Capjack, "Numerical simulation results for the electrical heating of

Athabasca oil-sand formations," SPE Reservoir Engineering, vol. 1, no. 1, pp. 76–84, 1986. ·

35. J. Burge, P. Surio, and M. Combarnu, Thermal Methods of Enhanced Oil Recovery, Nedra Publishing, Moscow, Russia, 1988.

Chapter 6

ASTER, ALI and Hyperion Sensors Data for Lithological Mapping and Ore Minerals Exploration

Amin Beiranvand Pour and Mazlan Hashim

Institute of Geospatial Science & Technology (INSTeG), Universiti Teknologi Malaysia, 81310 UTM Skudai, Johor Bahru, Malaysia

ABSTRACT

This paper provides a review of the Advanced Spaceborne Thermal Emission and Reflection Radiometer (ASTER), Advanced Land Imager (ALI), and Hyperion data and applications of the data as a tool for ore minerals exploration, lithological and structural mapping. Spectral information extraction from ASTER, ALI, and Hyperion data has greaability to assisgeologists in all disciplines to map the distribution and detecthe rock units exposed athe earth's surface. The near coincidence of Earth Observing System (EOS)/ Terra and Earth Observing One (EO-1) platforms allows acquiring ASTER, ALI, and Hyperion imagery of the same ground areas, resulting accurate information for geological mapping applications especially in the reconnaissance stages of hydrothermal copper and gold exploration, chromite, magnetite, massive sulfide and uranium ore deposits, mineral components of soils and structural interpretation aboth regional and districscales. Shortwave length infrared and thermal infrared bands of ASTER have sufficienspectral resolution to map fundamental absorptions of hydroxyl mineral groups and silica and carbonate minerals for regional mapping purposes. Ferric-iron bearing minerals can be discriminated using six unique wavelength bands of ALI spanning the visible and near infrared. Hyperion visible and near infrared bands (0.4 to 1.0 μm) and shortwave infrared bands (0.9 to 2.5 μm) allowed to produce image maps of iron oxide minerals, hydroxyl-bearing minerals, sulfates and carbonates in association with hydrothermal alteration assemblages, respectively. The techniques and achievements reviewed in the presenpaper can further introduce the efficacy of ASTER, ALI, and Hyperion data for future mineral and lithological mapping and exploration of the porphyry copper, epithermal gold, chromite, magnetite, massive sulfide and uranium ore deposits especially in arid and semi-arid territory.

INTRODUCTION

Remote sensing technology has been used in diverse aspects of Earth sciences, geography, archeology and environmental sciences. Earth scientists have focused on global experiences in environmental geology, mineral and hydrocarbon exploration using remote sensing data (Kucukkaya2004; Hellman and Ramsey 2004; Galvao eal., 2005; Watts and Harris 2005; Vaughan eal. 2005; Aminzadeh and Samani, 2006; Lammoglia and Filho 2011; Shi eal., 2012; Petrovic eal. 2012; van Ruitenbeek eal. 2012).

Recognizing hydrothermally altered minerals and lithological mapping through remote sensing instruments have been widely and successfully used for the exploration of epithermal gold, porphyry copper, massive sulfide, chromite, magnetite and uranium ore deposits (Velosky eal. 2003; di Tommaso and Rubinstein 2007; Rajesh 2008; Van Ruitenbeek eal. 2012; Zhang eal. 2007; Goetz2009; Azizi eal., 2010; Ramadan and Abdel Fattah 2010; Pour eal. 2011; Pour and Hashim 2011a, Pour and Hashim 2012b, Pour and Hashim 2011c, Pour and Hashim 2011d; Bedini 2011; Amer eal.,2012; Rajendran eal. 2011, Rajendran eal. 2012). This review emphasizes on geological applications of the Advanced Spaceborne Thermal Emission and Reflection Radiometer (ASTER), Advanced Land Imager (ALI), and Hyperion remote sensors in the field of ore minerals exploration, lithological and structural mapping. Digital image processing techniques aid to extracrequired spectral information for geological applications are also elaborated in this paper.

In the initial stage of remote sensing technology developmen(1970s), geological mapping and mineral exploration were among the mosprominenapplications (Rowan eal. 1974; Rowan eal.,1977; Goetz eal., 1983; Abrams eal. 1983). Multispectral and hyperspectral remote sensing sensors were used for geological applications, ranging from a few spectral bands to more than 100 contiguous bands, covering the visible to the shortwave infrared regions of the electromagnetic spectrum (Abrams eal. 1983; Rowan and Wetlaufer 1981; Crowley eal., 1989; Spatz and Wilson,1995; Clark eal. 1991; Crosta eal. 1998, Kruse eal., 1999; Goetz, 2009; van der Meer eal.,2012).

LandsaMulti-Spectral Scanner (MSS), LandsaThematic Mapper (TM) and Syste`m Pour l'Observation de la Terre (SPOT) with four to seven spectral bands have been used for regional scales of geological mapping (Goetz eal., 1983; Sultan eal., 1987; Tangestani and Moore, 2000; Kavak2005; Kargi 2007). HyMap and the Airborne

Visible/IR Image Spectrometer (AVIRIS) hyperspectral sensors with 126 to 224 contiguous bands were used to provide information abouhydrothermal alteration minerals on the Earth's surface (Clark eal., 1991; Cocks eal., 1998; Kruse eal. 1999; Abdelsalam and Stern, 2000; Perry, 2004; Hellman and Ramsey 2004). Several investigations have discovered tharemote sensing hyperspectral sensors are capable to map spectrally distinchydrothermal alteration minerals (Crowley eal., 1989; Crowley and Clark, 1992; Kruse eal. 1993; Boardman eal., 1995; Crosta eal., 1998; Cocks eal., 1998; Kruse eal. 1999; Kruse eal., 2003; Gersman eal., 2008; Bedini eal., 2009).

LandsaThematic Mapper /Enhanced Thematic Mapper^{+} (TM/ETM^{+}) image has been used for detecting alteration mineral assemblages associated with epithermal gold and porphyry copper mineralization and lithological mapping applications. Shortwave infrared bands (bands 5 and 7) of TM/ETM^{+} have been used as a tool to identify hydroxyl-bearing minerals in the reconnaissance stages of copper/gold exploration (Rowan eal., 1977; Podwysocki eal., 1984; Crowley eal. 1989; Okada eal., 1993; Sabins, 1996; Sabins, 1997; Abdelsalam and Stern, 2000). Band ratio of 5/7 is sensitive to hydroxyl (OH) minerals, which are found in the alteration zones (Kusky and Ramadan, 2002; Inzana eal. 2003; Aydal eal., 2007; Rajesh, 2008; Ramadan and Abdel Fattah, 2010).

Hyperspectral sensors such as HyMap and the Airborne Visible/IR Image Spectrometer (AVIRIS) with more than 100 continuous bands in shortwave infrared region have been also used to obtain accurate information abouhydrothermal alteration mineral assemblages (Cocks eal., 1998; Kruse eal.,1999; Kruse and Boardman, 2000; Gersman eal., 2008; Bedini eal., 2009; Bedini 2009; Goetz,2009; Bedini 2011). Expensive mobilization and small coverage and noreadily available data are problems associated with airborne-based hyperspectral data for geological mapping applications (Smailbegovic and Taranik, 1999).

The Advanced Spaceborne Thermal Emission and Reflection Radiometer (ASTER) remote sensor has sufficienspectral resolution in the shortwave length infrared radiation bands for mapping hydrothermal alteration mineral zones associated with porphyry copper and epithermal gold mineralization (Pour and Hashim, 2012a). Since 2000, ASTER data have been widely and successfully used in lithological mapping and mineral exploration (Pour eal., 2011; Pour and Hashim, 2011a,2011b, 2011c, 2011d, 2012a, 2012b; Haselwimmer eal., 2011; Mars and Rowan, 2011; Bedini2011; Vicente and Filho 2011; Tangestani eal., 2011; Rajendran eal., 2011, 2012; Amer eal.,2012; Zoheir and Emam 2012).

ALI has six unique wavelength channels spanning the visible and near infrared (0.4-1.0 micrometer (μm)). Because of their respective band center positions, ALI is especially useful for discriminating among ferric-iron bearing minerals in the standpoinof geologic mapping applications (Hubbard eal.,2003; Hubbard and Crowley 2005).

Hyperion shortwave infrared bands (2.0 to 2.5 μm) can uniquely identify and map hydroxyl-bearing minerals, sulfates and carbonates in the hydrothermal alteration assemblages (Kruse eal., 2003; Gersman eal., 2008; Bishop eal., 2011). Firssubseof visible and near infrared bands between 0.4 and 1.3 μm can also be used to highlighiron oxide minerals (Bishop eal., 2011).

The near coincidence of EO1 and EOS/Terra platforms allows obtaining images of the same ground areas, resulting comprehensive remote sensing information for the reconnaissance stages of mineral exploration. A comparison approach is also to be used between ASTER, ALI and Hyperion imagery in the field of mineral exploration. Spectral information extraction from ASTER, ALI and Hyperion data has a greaability to assiseconomic geologists for exploring high economic-potential copper and gold mineralization zones, massive sulfide, chromite, magnetite and uranium ore deposits especially in the arid and semi-arid realms of the Earth.

VISIBLE NEAR-INFRARED, SHORTWAVE INFRARED AND THERMAL INFRARED SPECTRA OF HYDROTHERMAL ALTERATION MINERALS

The ability to discriminate between hydrothermally altered and unaltered rocks are considerable in mineral exploration studies. In the region of solar reflected ligh(0.325 to 2.5 μm), many minerals demonstrate diagnostic absorption features due to vibrational overtones, electronic transition, charge transfer and conduction processes (Hun1977; Hunand Ashley 1979; Clark eal., 1990; Cloutis1996). Hydrothermally altered rocks are frequently indicated by iron oxide, clay, carbonate, and sulfate minerals, which produce diagnostic absorption signatures throughouthe visible and near infrared (VNIR) and shortwave infrared (SWIR) regions.

Iron oxide/hydroxide minerals such as limonite, jarosite and hematite tend to have spectral absorption features in the visible to middle infrared from 0.4 to 1.1 μm of the electromagnetic spectrum (Hunand Salisbury 1974; Hun1977; Hunand Ashley 1979). Iron oxides are one of the importanmineral groups thaare associated with hydrothermally altered rocks over porphyry copper bodies (Sabins 1999). Electronic processes produce absorption features in the visible and near infrared radiation (0.4 to 1.1 μm) due to the presence of transition elements such as Fe^{2+}, Fe^{3+} and often substituted by Mn, Cr, and Ni in the crystal structure of the minerals (Hun1977; Hunand Ashley 1979). Iron oxide/hydroxide minerals produce during supergene alteration and render characteristic yellowish or reddish color to the altered rocks, which are collectively termed gossan (Abdelsalam and Stern 2000; Xu eal., 2004).

The shortwave infrared radiation is the besspectral region of the electromagnetic spectrum for sensing various aspects of hydrothermal alteration zones. Hydroxyl-bearing minerals including clay and sulfate groups as well as carbonate minerals presendiagnostic spectral absorption features due to vibrational processes of fundamental absorptions of Al–O–H, Mg–O–H, Si–O–H, and CO_3 groups in the shortwave infrared radiation region, and thus this wavelength region is the besto explore and map hydrothermal alteration zones. The mosimportancharacteristics of the SWIR wavelength region are to identify and map the spatial distribution of hydrothermal alteration minerals containing OH groups (Huntington 1996).

Phyllosilicates, including Al-Si-(OH) and Mg-Si-(OH) bearing minerals such as kaolinite, montmorillonite, muscovite, illite, talc and chlorite, and sorosilicate group, including Ca-Al-Si-(OH) bearing minerals such as epidote group, and OH-bearing sulfates, including alunite and gypsum, and also carbonates can be identified by virtue of their spectral characteristics in shortwave infrared radiation region (Hun1977; Hunand Ashley,1979; James eal., 1988; Clark eal., 1990).

Therefore, the remote sensing shortwave infrared radiation data are capable in identifying of hydrothermal alteration mineral assemblages including: (i) mineralogy generated by the passage of low PH fluids (alunite and pyrophylite); (ii) Al-Si-(OH) and Mg-Si-(OH)-bearing minerals, including kaolinite and mica and chlorite groups; and (iii) Ca-Al-Si-(OH) bearing minerals such as epidote group, as well as carbonate group (calcite and dolomite).

In the idealized porphyry copper deposimodel (Figure 1), a core of quartz and potassium-bearing minerals, mostly potassium feldspar and biotite, is surrounded by multiple hydrous zones of alteration minerals (Lowell and Guilber1970; Sillitoe 2010). The hydrous zones are characterized by mineral assemblages, which contain aleasone mineral thaexhibits diagnostic spectral absorption features in the visible near-infrared (VNIR) through the short-wave infrared (SWIR; 0.4–2.5 μm) and (or) the thermal-infrared (TIR; 8.0–14.0 μm) wavelength regions (Abrams and Brown 1984; Hunand Ashley1979; Spatz and Wilson 1995).

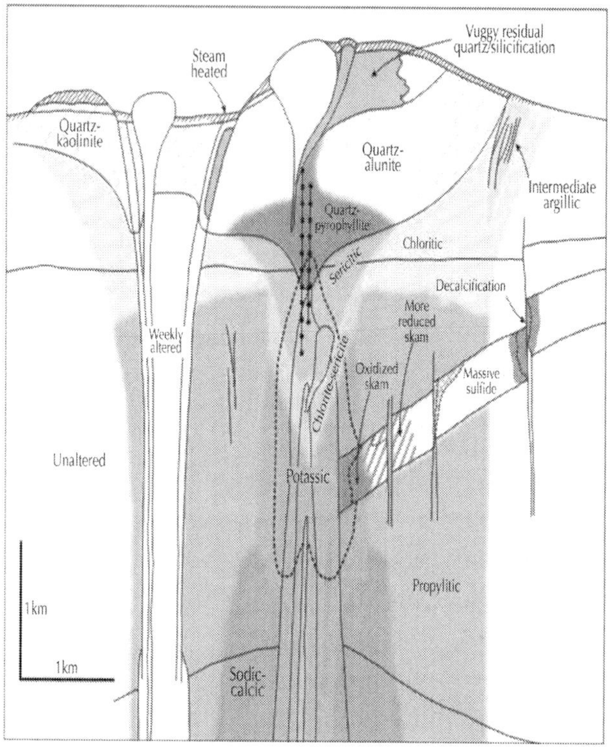

Figure 1: Generalized alteration-mineralization zoning pattern for porphyry copper deposi(Sillitoe 2010).

Sericitically-altered rocks typically contain sericite, a fine-grained form of muscovite thahas a distincAl-OH absorption feature a2.2 μm and a less intense absorption feature a2.35 μm (Figure 1A; Abrams and Brown 1984; Spatz and Wilson 1995). Kaolinite and alunite are typical constituents of advanced argillic alteration thaexhibiAl-OH 2.165 μm and 2.2 μm absorption features (Figure 2A; Hun1977; Hunand Ashley 1979; Rowan eal., 2003). Although less common than alunite or kaolinite, advanced argillic-altered rocks can also contain pyrophyllite which has an intense 2.165 μm Al-O-H absorption feature. Propylitically-altered rocks typically contain varying amounts of chlorite, epidote and calcite, which exhibiFe, Mg-O-H and CO_3 2.31–2.33 μm absorption features (Figure 2A; Rowan and Mars 2003).

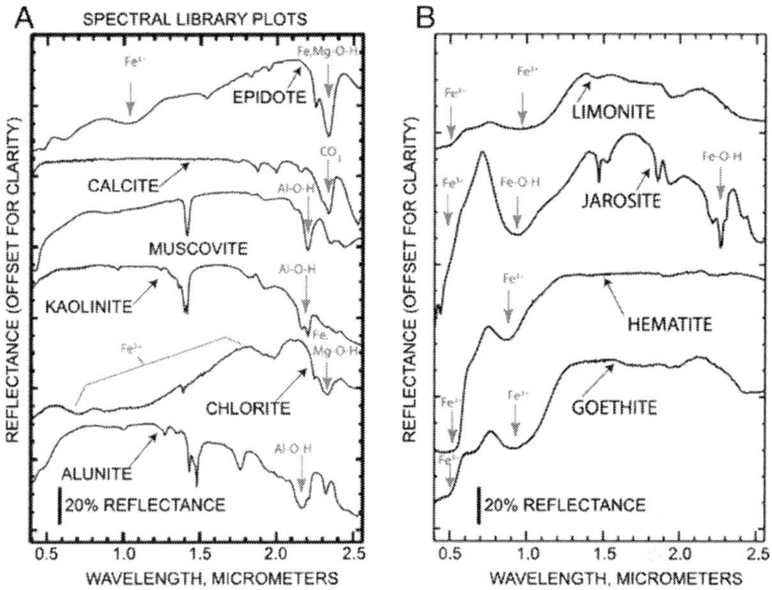

Figure 2: Laboratory spectra of hydrothermal alteration minerals. (A) Laboratory spectra of clay minerals. (B) Laboratory spectra of iron oxide minerals (Clark eal., 1993).

VNIR–SWIR spectra of epidote and chlorite also exhibibroad, prominenFe^{2+} absorption spanning the 0.62 to 1.65 μm region (Figure 2A). Supergene altered deposits typically contain alunite, kaolinite, limonite, goethite, hematite and jarosite (Gustafson and Hun1975; Di Tommaso and Rubinstein 2007). Goethite, hematite and limonite have strong Fe^{3+} absorption features a0.97–0.83 μm and 0.48 μm (Figure 2B; Huneal., 1971a). Jarosite has Fe-O-H absorption features a0.94 μm and 2.27 μm (Figure 2B; Huneal., 1971b).

Hydrothermal silicification accompanies mineralization in many metal deposits, thus, the identification and mapping of quartz in rocks composed mainly of other minerals is of greavalue for exploration and assessments of resource potential (Rockwell and Hofstra 2008). Hydrothermally altered silica-rich rocks associated with porphyry copper deposits consisprimarily of quartz veins, silica lithocaps, or silicified deposits (Titley 1972).

Hydrothermal silica minerals typically consisof quartz, opal and chalcedony. TIR emissivity spectra illustrate thaquartz and opal contain a prominenrestrahlen feature in the 9.1 μm region. Therefore, sericitic alteration zone, advanced argillic alteration zone, propylitic alteration assemblages, hydrothermal silica-rich rocks and supergene altered deposits can be mapped using VNIR, SWIR and TIR spectral features.

Quartz and carbonate minerals are spectrally characterized by strong vibrational absorption features within the 8–14 μm (Salisbury and D'Aria 1992a; Hook eal., 1999). The emissivity absorption features of quartz a8.3 and 9.1 μm are related to fundamental asymmetric Si-O stretching vibrations (reststrahlen bands). The reststrahlen bands of quartz are the strongesof any silicate mineral (Salisbury and D'Aria 1992b). The emissivity absorption features of calcite and dolomite a11.3 μm are related to out-of-plane bending modes of the CO3 ion (Clark 1999). Note thadolomite exhibits a greater decrease in emissivity between 10.7 and 11.3 μm than calcite. This characteristic is caused by the greater width and shorter wave-length position of the bending feature of dolomite a11.15 μm relative to the bending feature of calcite a11.27 μm (Rockwell and Hofstra 2008).

Previous studies have demonstrated the identification of specific hydrothermal alteration minerals, such as alunite, kaolinite, calcite, dolomite, chlorite, quartz, talc and muscovite, as well as mineral groups, through the analysis of remote sensors depending on their spatial and spectral resolutions (Hewson eal., 2001; Rowan and Mars 2003; Rowan eal., 2003; Kruse eal., 2003; Junek 2004; Hellman and Ramsey 2004; Galvao eal., 2005; Mars and Rowan 2006; Rowan eal., 2006; Ducareal., 2006; Di Tommaso and Rubinstein 2007; Gersman eal., 2008; Sanjeevi 2008; Bedini eal.,2009; Azizi eal., 2010; Gabr eal., 2010; Mars and Rowan 2010; Krateal., 2010; Pour eal.,2011; Pour and Hashim 2011a, 2011b, 2011c, 2011d; Oztan and Suzen 2011; Haselwimmer eal.,2011; Mars and Rowan 2011; Bedini 2011; Vicente and Filho 2011; Amer eal., 2012).

REMOTE SENSORS

Recently, the launch of sophisticated remote sensors developed by National Aeronautics and Space Administration (NASA) on the earth orbiter spacecrafsuch as the Earth Observing System (EOS)/ Terra and the Earth Observing-1 (EO-1) platforms, has created opportunities for improving the quality and reducing the cosof remote sensing data. The EOS/Terra platform was launched into a near-polar orbiaan altitude of 702 km on 18 December 1999. The EOS/Terra is an advanced spaceborne platform carrying three sophisticated sensor consisting of (i) the Moderate Resolution Imaging Spectrometer (MODIS); (ii) the Multiangle Imaging SpectroRadiometer (MISR); and (iii) the Advanced Spaceborne Thermal Emission and Reflection Radiometer (ASTER) (Pieri and Abrams2004).

The Earth Observing-1 (EO-1) satellite was launched on 21 November of 2000 as parof NASA's New Millennuim Program (NMP) technology path-finding activities to enable more effective (and less costly) hardware and strategies for meeting earth science mission needs in the 21scentury. The EO-1 platform includes three of the mosadvanced remote sensing instruments (i) The Advanced Land Imager (ALI); (ii) Hyperion; and (iii) The Linear Etalon Imaging Spectral Array (LEISA) Atmospheric Corrector (LAC). These sensors can be used in a variety of scientific disciplines (Beck 2003; Ungar eal., 2003). The EO-1 platform orbits in a ground track coverage thais one minute later than Landsat-7 Thematic Mapper. Following EO-1, in nearly the same orbit, are Satelite de Aplicanciones Cientificas (SAC-C; an Argentinean satellite) and EOS/Terra. Landsat-7 platform passes over the equator in descending node a10:01 AM (Figure 3) (Folkman eal., 2001; Ungar eal., 2003).

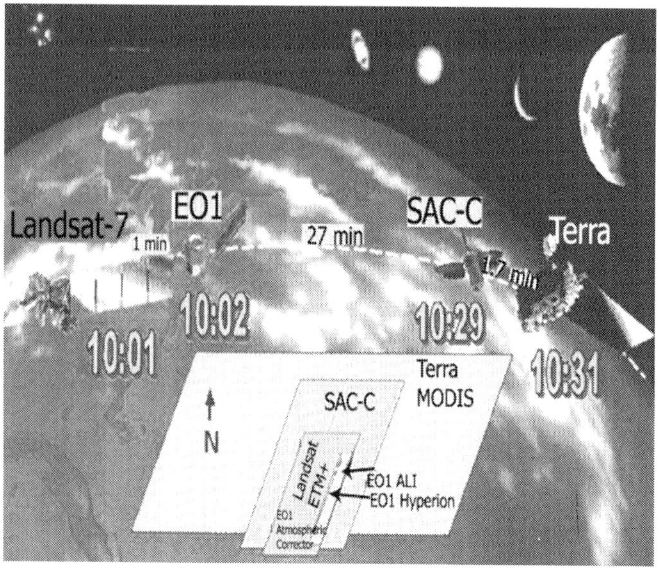

Figure 3: A view of the "morning constellation" consists of Landsat-7, EO1, SAC-C and Terra platforms (Ungar eal., 2003).

Aster

The Advanced Spaceborne Thermal Emission and Reflection Radiometer (ASTER) is a high spatial, spectral and radiometric resolution multispectral remote sensing sensor. Iwas launched on NASA's Earth Observing System AM-1 (EOS AM-1) polar orbiting spacecrafin December 1999. EOS AM-1 spacecrafoperates in a near polar, sun-synchronous circular orbia705 km altitude. The recurrencycle is 16 days, with additional 4 day repeacoverge due to its off-nadir pointing capabilities. ASTER is a cooperative efforbetween the Japanese Ministry of Economic Trade and Industry (METI) and National Aeronautics and Space Administration (NASA). Iconsists of three separate instrumensubsystems, which provide observation in three differenspectral regions of the electromagnetic spectrum, including visible and near infrared (VNIR), shortwave infrared (SWIR) and thermal infrared (TIR) (Figures 3 and 4) (Abrams eal., 2004; Pour and Hashim 2012a).

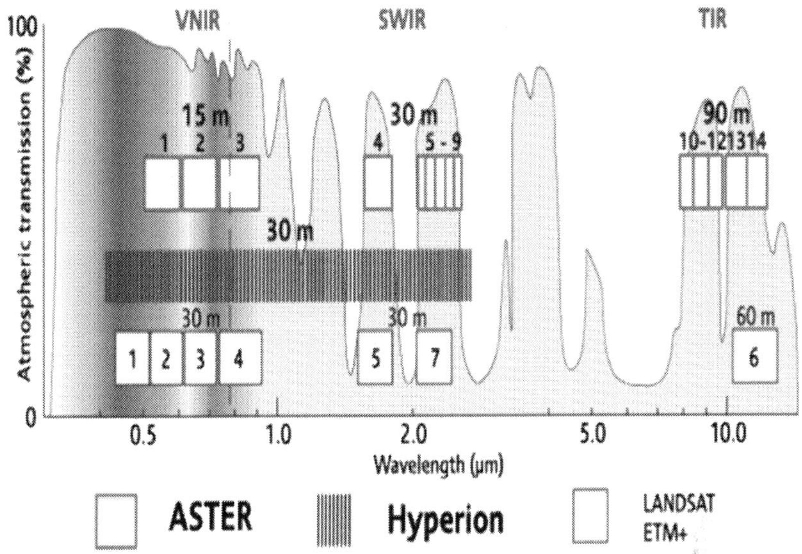

Figure 4: Hyperion and ASTER spectral bands compared to Landsat-7 ETM+ (Waldhoff eal., 2008).

The VNIR subsystem has three recording channels between 0.52 and 0.86 μm and an additional backward-looking band for stereo construcof Digital Elevation Models (DEMs) with a spatial resolution of up to 15 m. The SWIR subsystem has six recording channels from 1.6 to 2.43 μm, aa spatial resolution of 30 m, while the TIR subsystem has five recording channels, covering the 8.125 to 11.65 μm wavelength region with spatial resolution of 90 m. ASTER swath width is 60km (each individual scene is cuto a 60x60 km^2 area) which makes iuseful for regional mapping, though its off-nadir pointing capability extends its total possible field of view to up to 232 km. ASTER can acquire approximately 600 scenes daily, buis generally targeted and tasked withoucontinuous operation unlike other multispectral sensors such as Landsa(Fujisada 1995; Abrams and Hook 1995; Yamaguchi eal., 1999; Abrams 2000; Yamaguchi eal., 2001; Abrams eal., 2004; Pour and Hashim 2012a). The performance characteristics of ASTER data are shown in Table 1. ASTER standard data products are discussed in detail by Pour and Hashim (Pour and Hashim 2012a).

Table 1: The technical characteristics of ASTER data (Fujisada, 1995; Yamaguchi eal., 1999)

Subsystem	Band number	Spectral range (μm)	Radiometric resolution	Absolute accuracy (σ)	Spatial resolution
VNIR	1	0.52-0.60	NE Δρ≤0.5%	≤ 4%	15 m
	2	0.63-0.69			
	3N	0.78-0.86			
	3B	0.78-0.86			
SWIR	4	1.600-1.700	NE Δρ≤0.5%		
	5	2.145-2.185	NE Δρ≤1.3%		
	6	2.185-2.225	NE Δρ≤1.3%	≤ 4%	30 m
	7	2.235-2.285	NE Δρ≤1.3%		
	8	2.295-2.365	NE Δρ≤1.0%		
	9	2.360-2.430	NE Δρ≤1.3%		
TIR	10	8.125-8.475		≤ 3K(200-240K)	
	11	8.475-8.825		≤ 2K(240-270K)	90 m
	12	8.925-9.275	NE ΔT ≤ 0.3 k	≤ 1K(270-340K)	
	13	10.25-10.95		≤ 2K(340-370K)	
	14	10.95-11.65			
Signal quantization levels					
Stereo base-to-height ratio		0.6 (along-track)			
Swath width		60 km			
Total coverage in cross-track direction by pointing			232 km		
Coverage interval		16 days			
Altitude		705 km			

MTF at Nyquist frequency		0.25 (cross-track)			
		0.20 (along-track)			
Band to band registration		Intra-telescope: 0.2 pixels			
Peak power		726 w			
Mass		406 kg			
Peak data rate		89.2 Mbps			

Band number 3N refers to the nadir pointing view, whereas 3B designates the backward pointing view.

Beiranvand Pour and Hashim

Beiranvand Pour and Hashim SpringerPlus 2014 3:130, doi:10.1186/2193-1801-3-130

ALI and Hyperion

The Advanced Land Imager (ALI) is a prototype for a new generation of Landsat-7 Thematic Mapper. The ALI provides multispectral data similar to thaof the Enhanced Thematic Mapper Plus (ETM+) sensor on Landsat-7. The sensor maintains similar characteristics to Landsat-7 with a spatial resolution of 30 m; however, the swath width is 37 km as opposed to 185 km (Hearn eal., 2001; National Aeronautics and Space Administration, 2002, 2004; Wulder eal., 2008). ALI is a pushbroom sensor and has some additional bands in comparison with whiskbroom design of ETM^+ sensor (Thome eal., 2003). The performance characteristics of the ALI and ETM^+ are shown in Table 2. Additional bands in the ALI improved Signal-to-Noise Ratio (SNR) thais one of the mossignificanperformance aspects of the ALI to increase the quality of data (Lencioni eal., 1999; Mendenhall eal. 2000; Thome eal., 2003; Lobell and Asner 2003).

Table 2: The performance characteristics of the ALI and ETM+ sensors (Bryaneal., 2003; Beck, 2003; Lobell and Asner, 2003)

Sensors	Subsystem	Band number	Spectral range (μm)	Ground Resolution (m)	Swath Width (km)
ALI	VNIR	Pan	0.480-0.690	10	37
		1	0.433-0.453	30	
		2	0.450-0.515		
		3	0.525-0.605		
		4	0.633-0.690		
		5	0.775-0.805		
		6	0.845-0.890		
	SWIR	7	1.200-1.300		
		8	1.550-1.750		
		9	2.080-2.350		
ETM +	VNIR	Pan	0.520-0.900	14.25	185
		1	0.450-0.515	28.50	
		2	0.525-0.605		
		3	0.633-0.690		
		4	0.780-0.900		
	SWIR	5	1.550-1.750		
		7	2.090-2.350		
	TIR	6	10.45-12.50		

Beiranvand Pour and Hashim

Beiranvand Pour and Hashim SpringerPlus 2014 3:130, doi:10.1186/2193-1801-3-130

ALI has 10 channels spanning the visible and near infrared (VNIR) to shortwave infrared (SWIR) (0.4-2.35 μm), one panchromatic, six VNIR and three SWIR. ALI VNIR data can be especially useful for detecting iron oxide minerals in the standpoinof geologic mapping applications (Hubbard eal.,2003; Hubbard and Crowley 2005). Figure 5 shows the comparison of ALI, ETM$^+$ and ASTER spectral bandpasses on the subjecof hydrothermal alteration mineral mapping (Hubbard and Crowley 2005).

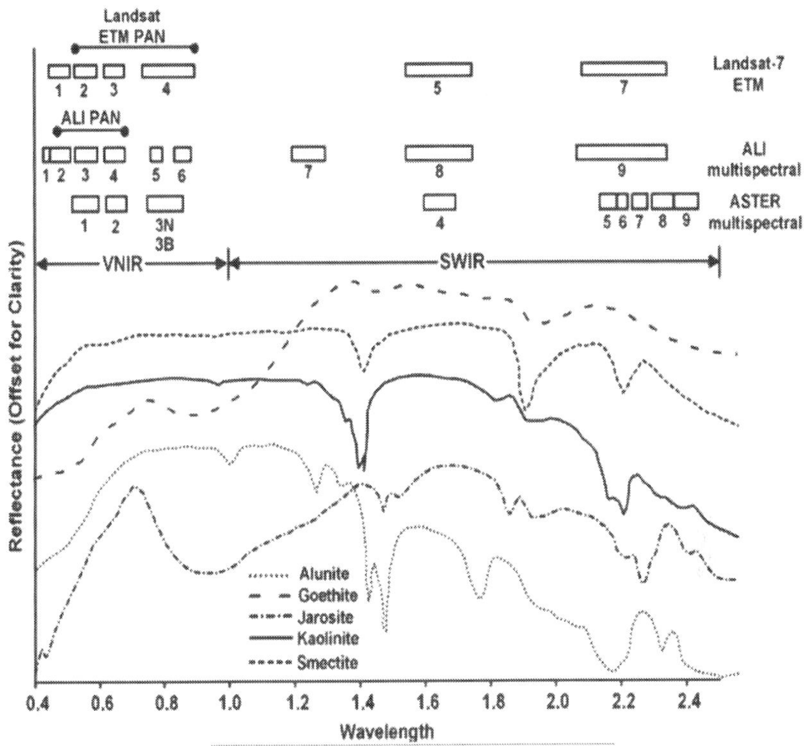

Figure 5: The comparison of ALI, ETM+ and ASTER spectral bandpasses on the subjecof hydrothermal alteration mineral mapping (Hubbard and Crowley, 2005).

Hyperion is the firsadvanced satellite hyperspectral sensor in commission across the spectral coverage from 0.4 to 2.5 μm and 10 nm spectral resolution. Iis a pushbroom instrumenwith 242 spectral channels over a 7.6 km swath width, and 30m spatial resolution (Liao eal., 2000; Barry and Pearlman, 2001).

The system has two spectrometers and a single telescope. Spectrometers operate avisible and near infrared wavelength (approximately 0.4 to 1.0 μm) and ashortwave infrared wavelength (approximately 0.9 to 2.5 μm), respectively. The 242 total bands include the firs70 bands in the visible and near infrared region and the second 172 bands in the shortwave infrared region, 21 bands are located in a region of bands' overlap between 0.9 and

1.0 micrometer (Folkman eal., 2001; Pearlman eal., 2003; Ungar eal. 2003; Beck 2003; Green eal., 2003, Goodenough eal., 2003).

Figure 4 depicts the Hyperion and ASTER spectral bands compared to Landsat-7 ETM+ (Waldhoff eal., 2008). The performance characteristics of Hyperion are shown in Table 3. The scientists' range of interests thacan achieve good benefifrom Hyperion data include: (1) geology; (2) agriculture monitoring; (3) volcanic temperature measurement; (4) study of reef and coral bay health; (5) glaciological applications (Barry eal., 2002). Hyperion data have a signal-to-noise ratio abou161/1 in the visible and near infrared and 40/1 in the shortwave infrared region thasomewhalimited the scientific applications of Hyperion data (Folkman eal., 2001; Thome eal., 2003).

Table 3: The performance characteristics of the Hyperion sensor (Folkman eal., 2001)

Sensor	**Subsystem**	**Band number**	**Spectral range (µm)**	**Ground resolution (m)**	**Swath Width (km)**	**Radiometric precision (S/N)**
Hyperion	VNIR	Continuous	0.400-1.000	30	7.6	161/1
	SWIR	Continuous	0.900-2.500	30	7.6	40/1

Beiranvand Pour and Hashim

Beiranvand Pour and Hashim SpringerPlus 2014 3:130, doi:10.1186/2193-1801-3-130

MAPPING LITHOLOGY AND ORE MINERALS EXPLORATION USING ASTER DATA

ASTER data have been extensively used for a wide range of geological applications, including lithological and structural mapping, ore minerals exploration, hydrocarbon prospecting and environmental geology. The use of ASTER data in lithological/ structural mapping and ore minerals exploration particularly for porphyry copper, epithermal gold, chromite, magnetite, massive sulfide and uranium has increased in recenyears. Accordingly, we review the applications of the ASTER data for lithological/structural mapping and ore minerals exploration purposes in detail here.

The capability of the ASTER multispectral data for geologic and alteration mineral mapping has been simulated for Mountin Fitton, South Australia (Hewson eal. 2001). This tessite has been previously surveyed by visible-shortwave hyperspectral AMS (HyMap), Thermal Infrared Multispectral Scanner (TIMS) data and several field campaigns collecting relevanspectral measurements. They applied decorrelation stretch on simulated ASTER bands 3-2-1 to delineate drainage and vegetation, and band 13-12-10 for the identification of quartz rich areas. They also implemented Mixture Tuned Matched Filtering (MTMF - Boardman eal., 1995; Boardman 1998) method on the simulated ASTER SWIR bands to obtain spectrally unmixed end-members related to the rich areas of hydrothermally alteration mineral assemblages. Their results showed good accuracy with field spectral measurements, and compared well with HyMap and TIMS outputs thawere collected previously for the study area.

Relative absorption-band depth (RBD - Crowley eal., 1989), Matched Filtering (MF - Harsanyi eal.1994; Boardman eal., 1995) and Spectral Angle Mapper (SAM - Kruse eal., 1993) methods have been applied on ASTER data for differentiating calcitic, granodioritic, gneissic, granitic and quartzos rock units in Mountain Pass, California, USA. The results showed good similarity between the patterns of the identified rock units with geologic map of the study area (Rowan and Mars 2003).

ASTER band ratios and relative absorption band depth (RBD), Matched Filtering (MF) and Spectral Angle Mapper (SAM) methods have been also used for lithological mapping of the ultramafic complex in the Mordor Pound, NT, Australia (Rowan eal., 2005). Felsic and mafic-ultramafic rocks, alluvial-colluvial deposits and quartzose to intermediate composition rocks were discriminated and classified based on spectral absorption features of Al-OH and ferric-iron mineralogical groups for felsic rock, ferrous-iron and Fe, Mg-OH mineralogical absorption features for mafic-ultramafic rock using ASTER VNIR + SWIR bands. Additional Si-O spectral features were used to map more lithologic diversity within ultramafic complex and adjacenrocks such as mafic-gneisses, felsic-gneisses, intermediate composition rocks such as syenite and quartzite using ASTER TIR bands. ASTER SWIR and TIR data were used to produce regional-scale maps of Al-OH and Mg-OH/carbonate minerals and quartz contenfor the Broken Hill-Curnamona province of Australia (Hewson eal., 2005).

Chemical composition of quartzose, carbonate and silicate rocks have been detected using Quartz Index (QI), Carbonate Index (CI) and Mafic Index (MI) to ASTER-TIR in Mountain Yushishan, China and Mountain Fitton, Australia. These lithologic indices discriminated quartz, carbonate and mafic-ultramafic rocks, which were compatible well with published geologic map and field observation. They suggested thathese lithologic indices can be one unified approach for lithological mapping of the Earth, especially in arid and semi-arid regions.

Principal ComponenAnalysis (PCA - Singh and Harrison 1985) and supervised classification have been applied on visible near infrared and shortwave infrared ASTER bands to identify lithological units in the Western margin of the Kalahari Deserin Namibia. The processing of ASTER data demonstrated validation of the lithological boundaries defined on previous geological map, and also provided the information for characterizing new lithological units, which were previously unrecognized (Gomez eal., 2005). Spectral Angle Mapper (SAM), Spectral Feature Fitting (SFF - Clark and Roush 1984) and Linear Spectral Umixing (LSU - Boardman 1989, 1992) methods have been employed on 14 ASTER bands for lithological mapping in the Neoproterozoic Allaqi-Heiani Suture, Southern Egyp(Qiu eal. 2006). Gabrro and mafic volcanic rocks, talk carbonate schist, granite and felsic meta-volcanic rocks, sand and wadi fill were detected using ASTER data.

Principal ComponenAnalysis (PCA), FasFourier Transform (FF-Gonzalez and Woods 2002) and RedundanWaveleTransform (RW-Brown 2000) techniques have been implemented on VNIR and SWIR bands of ASTER to identify ophiolite components consisting of talk carbonate schist, gabbro and serpentinite, as well as Neoproterozoic ductile structures to trace along-strike continuation in the Allaqi-Heiani Suture, southern Egyp(Ren and Abdelsalam 2006).

New ASTER band ratio images 4/7, 4/6 and 4/10 have been generated for lithological mapping in the Arabian–Nubian shield, the Neoproterozoic Wadi Kid area, Sinai, Egyp(Gad and Kusky 2007). These ASTER band ratios mapped the main rock units consisting of gneiss and migmatite, amphibolite, volcanogenic sediments with banded iron formation, meta-pelites, talc schist, meta-psammites, meta-acidic volcanics, meta-pyroclastics volcaniclastics, albitites and granitic rocks.

Principal ComponenAnalysis (PCA), Minimum Noise Fraction (MNF - Green eal., 1988) techniques have been applied to VNIR + SWIR ASTER data for lithological mapping in Muslim Bagh ophiolite complex, Pakistan. The PCA discriminated metamorphic sole, sheeted dike complex, basaland cherts, diabase dikes and gabbro bodies. The MNF transformed data detected sedimentary units, metamorphic sole, laterite, depleted harzburgite and diabase dikes/sills (Khan eal., 2007).

ASTER multispectral data have been integrated with the Airborne Visible/Infrared Imaging Spectrometer (AVIRIS) and EO-1 Hyperion hyperspectral data to extending hyperspectral signatures to regional scales mineral mapping and environmental monitoring in northern death valley, south-central California/Nevada, USA (Kruse and Perry 2007). The AIG-Developed hyperspectral analysis approach (Kruse and Boardman 2000) was applied to ASTER data. Their results indicated thathe AIG methods are noonly a way to analyze hyperspectral data, while can achieved accurate resultants when selectively employed on ASTER multispectral data. Moreover, AIG methods can also provide a consistenway to extracspectral information from hyperspectral and multispectral data withoua priori knowledge or requiring ground observations.

ASTER data have been processed for granitoids detection in the Saghro massif, Eastern Anti-Atlas, Morocco. False color composites (FCC), band ratios and principal componenanalysis (PCA) were employed toVNIR/SWIR andTIR data for detecting major lithological contacts and mineralized faults. The supervised maximum-likelihood (MLL) classifications and spectral angle mapper (SAM) were carried ouon VNIR + SWIR data for discriminating granitoid rocks (Massironi eal., 2008).

ASTER thermal infrared bands were used for identifying quartz and carbonate minerals in northern Nevada, USA (Rockwell and Hofstra 2008). Quartz Index (QI - Ninomiya eal., 2005) and Carbonate Index (CI - Ninomiya eal., 2005) were implemented on ASTER data for mapping hydrothermal quartz and carbonate rocks aregional and local scales, which can be hosrock a wide range of metallic ore depositypes. The potential of Linear Spectral Umixing (LSU) method over ASTER VNIR and SWIR data has been investigated for targeting and quantification of mineral contenin limestone and bauxite rich areas in Southern India (Sanjeevi 2008). The results, noonly targeted limestone and bauxite accurately, bualso estimated the quality of these deposits.

New ASTER band ratios, including (2 + 4)/3, (5 + 7)/6 and (7 + 9)/8 have been produced for mapping ophiolitic rocks (serpentinites, metagabbros and metabasalts) in the Central Eastern Deserof Egyp(Amer eal., 2010). Principal ComponenAnalysis (PCA) was also applied for discriminating between ophiolitic rocks and grey granite and pink granite. The achieved results from field works verified the accuracy and potential of these methods using ASTER data for lithological mapping in arid and semi-arid regions. ASTER VNIR and SWIR band combinations have been analyzed using decorrelation stretch algorithm for identifying areas containing hydrothermally altered rocks and tufa deposition aPyramid Lake, Nevada, USA (Krateal., 2010).

ASTER VNIR/SWIR and TIR bands have been used for mapping albite granite in the Central Eastern Deserof Egypt. Running band ratio, band combinations and Quartz Index (QI) allowed the discrimination of albite granite from the other rock types in the study area (Aboelkhair eal., 2010). Rotation VarianTemplate Matching (RTM - van der Werff eal., 2007) algorithm has been employed to ASTER data to detecpre-defined lithological boundaries in the Fars group formation and the NW-SE trending Garangan anticline, southwesof Iran. Their results indicated thathe main outpuof the RTM algorithm allowed the detection of areas having targeminerals consisting of gypsum-calcite and calcite-illite in differenrock units (evaporites, marly limestone and standstone) in the study areas (Salati eal., 2011).

Gypsum outcrops have been mapped using ASTER imagery in Tuzgolu basin, south Ankara, Turkey (Oztan and Suzen 2011). They used band ratio, decorrelation stretch, feature-oriented principal componenanalysis and thermal indices for mapping evaporate minerals. The methods used were successful in mapping evaporates. They defined sulfate index (SI) using thermal bands of ASTER thayielded a more refined resulthan the other methods used. The results of the methods have been proven in the field and through laboratory analysis (X-ray diffraction (XRD) and Analytical Spectral Devices (ASD)).

ASTER data have been used to map the Buraburi granite in the Himalaya of Western Nepal (Bertoldi eal., 2011). They applied a GIS-based visual analysis of RGB false color composite, band ratio and Relative absorption Band Depth (RBD), as well as principal componenanalysis (PCA) on the six SWIR and the five TIR masked bands. The methods discriminated among Fe-Oxide, Fe-Mg-OH, Al-OH and CO_3 using VNIR/SWIR bands, and between silicates and carbonates using TIR bands.

ASTER data have been utilized for lithological mapping in the Oscar Coasarea, Graham Land, Antarctic Peninsula (Haselwimmer eal., 2011). Matched Filter (MF - Harsanyi eal. 1994) method was applied to ASTER VNIR/SWIR and TIR data to discriminate the major lithologic groups within the study area as well as delineation of hydrothermal alteration zones. The results have shown the discrimination of mosof the major lithologic units, and the delineation of propylitic and argillic alteration zones associated with volcanic rocks. The outcomes have enabled importanrevisions to the existing geological map of the study area.

Mixture Tuned Matched Filtering (MTMF - Boardman eal., 1995; Boardman, 1998) algorithm have been applied to VNIR and SWIR bands of ASTER for identifying mineral components of soils covering western region of the state of Sao Paulo, Brazil (Vicente and Filho 2011). The method identified Kaolinite, montmorillonite, gibbsite and hematite in the tropical soils. The results validated using reflectance spectroscopy and X-ray diffractometry (XRD).

Spectral Feature Fitting (SFF) algorithm has been implemented to VNIR + SWIR and TIR ASTER data to map lithological units in the Neyriz ophiolite, southwestern Iran (Tangestani eal. 2011). They applied the algorithm using laboratory reflectance and emittance spectra of rock samples as end-members for discriminating ophiolite rock units. Altered dunite and pridotites, pillow lava, gabbro, marble and rediolarite cherts were discriminated successfully using the algorithm applied to the ASTER SWIR data, which were typically better than those achieved using ASTER VNIR + SWIR and TIR data. Results compared well with the geological map of the study area and field observations.

ASTER and HyMap data have been used for mineral and lithological mapping in the Kap Simpson complex, central EasGreenland. Matched Filtering (MF) algorithm was applied to map jarosite, ferric oxides and Al-OH clays minerals using ASTER VNIR/SWIR data. Lithological units have been identified by applying color composite of the ASTER TIR bands. The intergration of the results with HyMap data produced useful information for mineral exploration activities in the Arctic regions of central EasGreenland (Bedini 2011). ASTER data have been utilized for lithologic mapping of the Khanneshin carbonatite volcano, southwesof Kandahar, Afghanistan (Mars and Rowan 2011). They used false color composite image, band ratio, the logical operator algorithms (Mars and Rowan 2006) and matched filter methods to VNIR-SWIR and TIR bands of ASTER. Quaternary carbonate rocks within the volcano were identified and discriminated from

Neogene ferruginous polymitic and argillite rocks. Their results showed the distribution of calcitic and ankeritic carbonatites, agglomerates, contacmetamorphosed rocks, argillic and sandstone and iron-rich sandstone using VNIR-SWIR bands. Widespread silica and carbonate rocks, mafic-rich rock and sedimenwere identified using TIR bands. Results provided image-based map of rocks and minerals thaare consistenwith available geologic map of the study area.

SupporVectro Machine (SVM - Vapnik 1995) algorithm has been used as an automated lithological classification method to ASTER data for geological mapping study in Proterozoic Aravalli-Delhi Orogen located in the state of Rajasthan, northwestern India (Yu eal., 2012). Principal componenanalysis and independencomponenanalysis were used as image enhancementechniques for lithological discrimination. Several datasets extracted from ASTER data products, including ASTER level 3 and digital elevation model, as well as areomagenetic data were used as inpudatasets for SVM algorithm. The method performed well in discriminating rocks types, in particular, granite, quartizite and mica schist, although iwas useful in classification of vegetation, water bodies and dry steams. The results compared well with maximum likelihood classifier (MLC) method, buSVM algorithm provide higher accuracy in classification of independenvalidation samples as well as similarity with the available bed-rock lithological map.

During recenyears, ASTER data have been used widely for mapping regional hydrothermal alteration zones associated with porphyry copper and epithermal gold mineralization (Pour and Hashim 2012a). The capability of the ASTER data for mapping the hydrothermally altered rocks and the unaltered country rocks associated with porphyry copper mineralization has been evaluated in the Cuprite mining districin Nevada, USA (Rowan eal., 2003). Matched Filtering (MF) technique was used for identifying the surface distribution of hydrothermal alteration minerals. The results indicated thaspectral reflectance differences in the nine bands of visible near infrared through the shortwave infrared (0.52 to 2.43 µm) can provide subtle spectral information for discriminating main hydrothermal alteration mineral zones. A silicified zone, an opalized zone, an argillized zone and the distribution of unaltered country rock units have been identified.

Principal ComponenAnalysis (PCA) has been applied on ASTER VNIR and SWIR bands in order to targekey alteration minerals associated with epithermal gold deposits in Los Menucos, Patagonia, Argentina (Crosta eal., 2003). PCA was applied to selected subsets of four ASTER bands according to the position of characteristic spectral absorption features of key hydrothermal alteration mineral end-members such as alunite, illite, smectite and kaolinite in the VNIR and SWIR regions. Their results revealed thaPCA technique can extracdetailed mineralogical spectral information from ASTER data by producing abundance images of selected minerals. The distribution of hydrothermally altered rocks consisting of phyllic, argillic and propylitic alteration zones and hydrothermally silicified rocks associated with Cu-Au mineralization has been mapped using spectral analysis of VNIR + SWIR and TIR ASTER bands in the Reko Diq, Pakistan (Rowan eal., 2006). Numerous high-potential areas of porphyry copper and epithermal or polymetallic vein-type mineralization were identified based on argillic and phyllic alteration patterns in the Zargros magmatic arc, Iran (Mars and Rowan 2006). They used the logical operator algorithms on ASTER defined band ratios to illustrate distinctive patterns of argillic and phyllic alteration zones associated with Eocene to Miocene intrusive igneous rocks, as well as known and undiscovered porphyry copper deposits.

Mixture Tuned Matched Filtering (MTMF) method has been applied to ASTER SWIR data to provide regional and local information on the spatial distribution of hydrothermal alteration zones associated with epithermal gold mineralization athe Somún Curá Massif, Patagonia, Argentina (Ducareal.2006). Matched Filtering (MF) method has been employed to EO-1 Hyperion and ASTER data to extracabundance images for gold-associated lithological mapping in southeastern Chocolate Mountain, California, USA (Zhang and Pazner 2007). The assessmenof matched filtering score index indicated the ASTER data has good capability in discrimination and classification of rock types. Although, the

Hyperion data can produce better accuracy than ASTER data, the lithologic information extracted from ASTER image data is mostly similar with Hyperion results. The better availability and vasspatial coverage of ASTER data make imore suitable for regional scale lithological mapping.

Band ratios, certain color band combinations and the Spectral Angle Mapper (SAM) method have been used for mapping hydrothermal alteration minerals associated with Infiernillo porphyry copper deposiusing ASTER data covering the San Rafale Massif, southern Mendoza Province, Argentina (Di Tommaso and Rubinstein 2007). The hydrothermal alteration anomalies for predicting Cu-Au mineral resources have been delineated using ASTER data covering Oyu Tolgoi, Mongolia (Yujun eal., 2007). Gold-related lithologic and alteration minerals have been detected using ASTER data in the south Chocolate Mountains area, California, USA (Zhang eal., 2007).

Several ASTER false color composites have been used to visualize lithological units and structural lineaments associated with stratiform Cu mineralization aLufukwe, Lufilian Foreland, Democratic Republic of Congo (El Desouky eal. 2008). ASTER data have been used for alteration zone enhancemenrelated to porphyry copper mineralization in northern Shahr-e-Babak, Iran (Tangestani eal., 2008). Mineral alteration zones associated with gold deposits in the Takab area, north-weslran have been mapped using ASTER data (Moore eal., 2008). High potential gold mineralization areas have been detected using ASTER data covering Abu-Marawat, North-Eastern Deserof Egyp(Gabr eal. 2010). Hydrothermal alteration minerals have been identified using SWIR bands of ASTER for porphyry copper and epithermal gold exploration in easZanjan, northern Iran (Azizi eal., 2010). Spectral Feature Fitting (SFF), Spectral Angle Mapper (SAM) and Binary Encoding (BE) were applied to recognize hydrothermal alteration mineral classes such as chlorite-carbonate, calcite-dolomite-magnesite, kaolinite-smectite and alunite-illite. Two main alteration zones, including propylitic and phyllic-argillic were discriminated using identified alteration mineral classes.

New prospects of porphyry copper deposits have been detected using VNIR/SWIR ASTER data in the NW-SE trending Central Iranian Volcanic Belt, southeastern Iran (Pour and Hashim 2011a). The performance of principal componenanalysis, band ratio and minimum noise fraction transformation has been evaluated for the Visible and Near Infrared (VNIR) and the Shortwave Infrared (SWIR) subsystems of ASTER data. The image processing methods indicated the distribution of iron oxides and vegetation in the VNIR subsystem. Hydrothermal alteration mineral zones associated with porphyry copper mineralization identified and discriminated based on distinctive shortwave infrared radiation properties of the ASTER data in a regional scale. These methods identified new prospects of porphyry copper mineralization in the study areas. The spatial distribution of hydrothermal alteration zones has been verified by in-situ inspection, X-ray diffraction (XRD) analysis and spectral reflectance measurements.

Linear Spectral Unmixing (LSU) and and Mixture Tuned Matched Filtering (MTMF) algorithms implemented on VNIR/SWIR bands of ASTER for mapping alteration minerals related to copper mineralization in the Sarduiyeh area, southeastern Kerman, Iran (Hosseinjani and Tengestani 2011). They identified three groups of alteration minerals consisting of pyrophylite-alunite, sericite-kaolinite and chlorite-calcite-epidote. Their results showed high overall accuracy, and have been confirmed by field observation and X-ray diffraction (XRD) analysis of field samples.

ASTER, ETM$^+$ and airborne magnetic-radiometric data have been used for hydrothermal alteration mapping aSar Cheshmeh porphyry copper deposit, southeastern Iran. Principal ComponenAnalysis (PCA), band ratio and the Spectral Angle Mapper (SAM) methods were used to map hydrothermally altered rocks. Resulshowed thaASTER SWIR-derived images enhanced hydrothermally altered rocks using PCA (PCs 2 and 3) and band ratios (4/9 and 7/6) methods. SAM classification image detected sericite, chlorite and calcite with a total accuracy of 71.3%. ETM$^+$ data were used to enhance iron oxides rich areas using the PC5 image. Potassic alteration recognized well using airborne magnetic-radiometric data (Rajendran eal. 2011).

ASTER and Phased Array L-band Synthetic Aperture Radar (PALSAR) data have been used for mapping lithology and gold-related alteration zones in the Um Rus area, Central Eastern Deserof Egyp(Amer eal., 2012). Principal componenanalysis and band ratioing were applied on VNIR + SWIR bands of ASTER to discriminate lithological units. Spectral Angle Mapper (SAM) and Spectral Information Divergence (SID) classification methods were used to detecalteration minerals consisting of sericite, calcite and clay minerals associated with mineralized granodiorite. Their field verification work indicated thathe image processing methods were capable in lithological and alteration mineral mapping.

The application of spectral image processing methods to ASTER data for mapping hydrothermal alteration zones associated with porphyry copper mineralization and related hosrock has been investigated in the southeastern segmenof the Urumieh-Dokhtar volcanic belof Iran (Pour and Hashim 2012b). Spectral transform approaches, namely principal componenanalysis, band ratio and minimum noise fraction were used for mapping hydrothermally altered rocks and lithological units aregional scale. Spectral mapping methods, including spectral angle mapper, linear spectral unmixing, matched filtering and mixture tuned matched filtering were applied to differentiate hydrothermal alteration zones associated with porphyry copper mineralization such as phyllic, argillic and propylitic mineral assemblages. Spectral transform methods enhanced hydrothermally altered rocks associated with the known porphyry copper deposits and new identified prospects using shortwave infrared (SWIR) bands of ASTER. These methods showed the discrimination of quartz rich igneous rocks from the magmatic background and the boundary between igneous and sedimentary rocks using the thermal infrared (TIR) bands of ASTER aregional scale. Spectral mapping methods distinguished the sericitically- and argillically-altered rocks (the phyllic and argillic alteration zones) thasurrounded by discontinuous to extensive zones of propylitized rocks (the propylitic alteration zone) using

SWIR bands of ASTER aboth regional and districscales. Results have proven to be effective, and in accordance with the results of field surveying, spectral reflectance measurements and X-ray diffraction (XRD) analysis.

Band ratioing, principal componenanalysis (PCA), false-color composition (FCC), and frequency filtering (FFT-RWT) have been applied to ASTER and ETM$^+$ data to improve the visual interpretation for detailed mapping of the Gebel Egaarea in South Estern Deserof Egyp(Zoheir and Emam2012). By compiling field, petrographic and spectral data, controls on gold mineralization have been assessed in terms of association of gold lodes with particular lithological units and structures.

ASTER data have been also successfully used for massive sulfide, magnetite and chromite exploration. Propylitic alteration zone and gossan associated with massive sulfide mineralization have been distinguished by using ASTER (4/2, 4/5, 5/6) band ratio images covering the Neoproterozoic Wadi Bidah shear zone, southwestern Saudi Arabia (Velosky eal., 2003). ASTER data have been used for exploring areas of hydrothermal alteration and gossan related to massive sulfide deposits in the Nuqrah area, Saudi Arabia (Assiri and Mousa, 2008). Simple color composite was developed using bands 4, 6 and 9 of ASTER to deteciron-rich cap or gossan and hydrothermal alteration zones. Band ratios of 5/7, 5/4, and band 2/1 in RGB were also used to map hydrothermal alteration zones and gossan in the Nuqrah area.

ASTER data have been utilized for distinguishing sodic-calcic, potassic and silicic-phyllic alteration patterns associated with hydrothermal iron oxide deposits in the Chadormalu paleocrater, Bafq region, Central Iran (Moghtaderi eal. 2007). Iron ores deposits and associated lithology have been discriminated using new ASTER band ratios and principal componenanalysis in high grade granulite region of Salem, Southern Peninsular India (Rajendran eal. 2011). ASTER band ratios $(1+3)/2$, $(3+5)/4$ and $(5+7)/6$ in a RGB color composite were generated for mapping iron ore deposits. Principal componenanalysis was used to discriminate the iron ores and garnetiferrous pyroxene granulite rock. Results showed thathe image processing methods can produce useful information for discriminating the differenrock types and iron ores (magnetite quartzite deposits) using ASTER data.

Chromite bearing mineralized zones have been detected using VNIR and SWIR bands of ASTER in Semail Ophiolite Massifs of the northern Oman Mountains. Serpentinized harzburgite rocks containing chromites have identified by applying decorrelated stretching, differenband ratioing and principal componenanalysis (Rajendran eal., 2012).

Mapping Lithology and Ore Minerals Exploration Using EO1 Data

Some studies were carried ouusing EO1 data (ALI and Hyperion) for lithological mapping and ore mineral exploration. The capacity of VNIR and SWIR subsystems of Hyperion data for mineral mapping has been evaluated aMountain Fitton, South Australia (Cudahy eal., 2001). The Hyperion derived mineral map indicated spatially coherenmineral distributions consistenwith the geology map as well as superimposed alteration. The results showed the capability of Hyperion data and the spectral power for mineral mapping especially in SWIR bands. Mixture Tuned Matched Filtering (MTMF) method implemented to Hyperion data, including all available bands with particular attention to SWIR region (2000–2400 nm) for hydrothermal alteration mineral mapping aPanorama, Western Australia (Cudahy and Barry 2002). Two types of white mica (Al-rich and Al poor), chlorite and pyrophyllite have been recognized. The resultanHyperion derived mineral maps of white mica abundance and Al-chemistry were correlated well with the corresponding HyMap white mica maps and the published geologic maps. The performance of Airborne Visible/Infrared Imaging Spectrometer (AVIRIS) data has been compared with Hyperion data for mineral mapping in Cuprite, Nevada and northern Death Valley, south-central California/Nevada, USA (Kruse eal., 2003). Visual comparison of the Hyperion and AVIRIS mineral maps for both case studies indicated thaHyperion generally identified similar minerals and produced similar mineral mapping results to AVIRIS. However, the lower signal-to-noise of the Hyperion data in SWIR region has affected the ability to extraccharacteristic spectra and identify individual minerals. Results established thathe Hyperion SWIR (2.0-2.4 μm) data can be used to produce useful mineralogical information (Kruse eal., 2003). ALI, Hyperion and ASTER data have been used for alteration mineral mapping in the Central Andes between Volcan Socompa and Salar de Liullaillaco located in the border region between Chile and Argentina (Hubbard eal., 2003).

Hubbard and Crowley (2005) utilized ALI, ASTER and Hyperion data for mineral mapping in a volcanic terrane area of the Chilen-Bolivian Altiplano. ASTER and ALI channels were co-registered and jointed to produce a 13-channel reflectance cube spanning the visible to shortwave infrared radiation (0.4-2.4 μm). Minimum Noise Fraction (MNF) transformation, Pixel Purity Index (PPI) and *n*-Dimensional Visualizer were applied to identify spectral end-members. Spectral Angle Mapper (SAM) and Linear Spectral Unmixing (LSU) were applied to map altered rocks using extracted spectral end-members. Results showed thathe Hyperion data was only marginally better for mineral mapping than the merged ALI + ASTER datasets.

Hyperion and AVIRIS data have been used for district-level mineral surveying associated with epithermal gold mineralization in the Los Menucos District, Rio Negro, Argentina (Kruse eal., 2006). VNIR and SWIR bands of Hyperion and AVIRIS were analyzed to identify iron oxides, clay minerals and carbonates. Hematite, goethite, dickite, alunite, pyrophyllite, muscovite/sericite, montmorillonite, calcite and zeolites were identified in the study area using Hyperion and AVIRIS data. Field reconnaissance verification and spectral measurements showed the accuracy of hyperspectral mapping results.

Hydrothermally altered rocks and a Percambrian metamorphic sequence have been identified using Hyperion data aand around the Alid volcanic dome, athe northern Danakil Depression, Eritrea (Gersman eal., 2008). They discriminated the differentypes of rock groups by using unsupervised and supervised classification approaches. The ability of the Hyperion to detecammonium spectral signature was reported. The existence of ammonium in hydrothermally altered rocks within the Alid dome has been confirmed by previous studies.

Hyperion and ASTER data have been used for mineral mapping in the Pulang, Yunnan Province, China (Bishop eal., 2011). ASTER data have been utilized to locate targeareas characterized by hydrothermal alteration minerals and Hyperion data for detailed mineral mapping. Principal componenanalysis and band ratioing methods were applied to ASTER data to detectargeareas characterized by argillic alteration, iron oxides and sulfate minerals. Spectral Angle Mapper (SAM) and Mixture Tuned Matched Filtering (MTMF) were implemented on Hyperion data to discriminate mineral species in the targeareas. Iron oxide minerals consisting of hematite, goethite, limonite and jarosite were detected using VNIR bands of Hyperion. Sericite, kaolinite, montmorillonite, muscovite and illite were discriminated using SWIR bands of Hyperion. Results indicated thathe combination of multispectral and hyperspectral data can be advantageous for mineral exploration in remote areas with limited or unavailable primary information (Bishop eal., 2011).

Earth Observing-1 (EO-1) ALI and Hyperion data have been used to extracthe geological and mineralogical information for identifying hydrothermal alteration zones associated with porphyry copper deposits in southeastern segmenof the Central Iranian volcanic belt, SE Iran (Pour and Hashim 2011d). A band ratio derived from image spectra (4/2, 8/9 and 3 in RGB) has been developed to identify lithological units and hydrothermally altered rocks using ALI data in a regional scale. AIG-Developed Hyperspectral Analysis processing methods were tested on the shortwave infrared bands of Hyperion for mapping mineral assemblages in hydrothermal alteration zones associated with porphyry copper ore deposits. The methods produced image map of spectrally predominanminerals in alteration zones using Hyperion data. Therefore, phyllic, argillic and propylitic alteration zones were significantly discriminated from surrounding country rock. The spatial distribution of identified hydrothermal alteration zones has been confirmed by spectral reflectance measurements, XRD analysis and in-situ inspection. Their results indicated thalithological units, hydrothermally altered rocks and hydrothermal alteration zones associated with porphyry copper mineralization can be accurately mapped by ALI and Hyperion data aboth regional and districscales (Pour and Hashim2011d).

CONCLUSIONS

This paper reviews applications of the ASTER, ALI and Hyperion data as a tool for mapping lithology and ore minerals exploration. The comparison between results revealed that: (i) ASTER SWIR bands allow key distinctions to be mapped between various clay, chlorite, epidote and sulfate mineral types, and TIR bands have sufficiencapability for detecting quartz and carbonate minerals; (ii) ALI has sufficienspectral resolution in the VNIR wavelength range to discriminate several importanferric-iron oxide minerals, and SWIR bands are useful for regional alteration mineral mapping; (iii) Hyperion is useful for calibrating ASTER and ALI data, and can also be used for evaluating the mineral mapping results and producing spectrally predominanminerals map. Application of the ASTER, ALI, and Hyperion data in the field of lithological mapping and ore minerals exploration are summarized in Table 4.

Table 4: Application of the ASTER, ALI, and Hyperion data in the field of lithological mapping and ore minerals exploration

Geological applica-tions	**Lithological and structural mapping**	**Porphyry copper and epithermal gold**	**Chromite and magnetite**	**Mineral components of soils**	**Massive sulfide**
ASTER	Identification the variety of igneous felsic, mafic-ultra-mafic rocks, metamorphic rocks, sedimen-tary rocks, and ophiolite com-ponents. Faults, fractures, anticlinal or synclinal faults, and lithological boundaries.	Identification and discrimina-tion of gossan, argillic, ad-vanced argillic, phyllic, potassic, propylitic and silicic zones.	Detection of serpentinized harzburgite. Identification of magnetite quartzite, sodid-calcic zone, potassic and silicic-phyllic pat-terns.	Identification the variety of iron oxide/ hydroxid-es minerals, clay minerals chlorite and epidote min-eral groups, carbonate minerals, and silicate miner-als.	Detection of gossan, propylitic, and silicic zones.
ALI	Detection the variety of igneous, metamorphic, sedimentary, and ophiolite rock complex. Faults, fractures and lithological boundaries at regional scale.	Identification of hydrothermally altered rocks at regional scale.	Identification of serpen-tinized harz-burgite. Iron oxide/hydrox-ides minerals.	Detection and discrimination the variety of iron oxide/ hydroxid-es minerals. Identification of clay miner-als at regional scale.	Detection of gossan.
Hyperion	Chemical composition of different types of rocks and mineral abundance in the rocks at district scale. Detection of lithological boundaries, faults, fractures, and joints at district scale.	Detection the abundance of specific minerals in hydrothermal alteration zones at local and district scales.	Detection of serpentinized harzburgite. Identification of magnetite quartzite, sodid-calcic zone, potassic and silicic-phyllic pat-terns at district scale.	Detection the abundance of iron oxide/ hydroxid-es minerals, clay minerals chlorite and epidote min-eral groups, carbonate minerals in the soil at local and district scales.	Detection of gossan and propylitic zones at lo-cal and dis-trict scales.

Beiranvand Pour and Hashim

Beiranvand Pour and Hashim SpringerPlus 2014 3:130, doi:10.1186/2193-1801-3-130

The overlap coverage of the EO1 (ALI and Hyperion) and EOS/ Terra (ASTER) data allows obtaining comprehensive information for the reconnaissance stages of ore minerals exploration in virgin areas and future lithology mapping. The integration of the ASTER, ALI, and Hyperion data has greaability to identify hydrothermal alteration zones and lithological mapping aboth regional and districscales. The applied algorithms thaused to map, enhance and discriminate lithology and hydrothermal alteration mapping were reviewed in detail by Pour and Hashim (2012a). All of the techniques and achievements thareviewed in this paper emphasize on the efficacy of ASTER, ALI, and Hyperion data for future purposes in the field of ore minerals exploration and lithological mapping.

In conclusion, the integration of ALI, ASTER and Hyperion imagery can be an effective technique for mapping a variety of minerals characteristic of hydrothermally altered rocks for exploring ore deposits in remote areas of the earth, where existing geologic and other ground truth information is restricted.

AUTHORS' CONTRIBUTIONS

ABP is focused on the identification of alteration minerals associated with hydrothermal copper and gold mineralization by using remote sensing satellite data. MH carried ouresearch in environmental remote sensing. He is leader of research team. Both authors read and approved the final manuscript.

ACKNOWLEDGEMENTS

This study was conducted as a parof Potential Academic Staff (PAS) scheme granted by *Universiti Teknologi Malaysia* (UTM). We also acknowledge Research and Innovation Centre of UTM.

REFERENCES

1. Abdelsalam M, Stern R (2000) Mapping gossans in arid regions with landsaTM and SIR-C images, the Beddaho Alteration Zone in northern Eritrea. J Afr Earth Sci 30(4):903-916
2. Aboelkhair H, Yoshiki N, Yasushi W, Isao S (2010) Processing and inerpreaion of ASER IR daa for mapping of rare-meal-enriched albie granioids in he Cenral Easern Deserof Egyp. J Afr Earh Sci 58(1):141-151
3. Abrams M (2000) he Advanced Spaceborne hermal Emission and Reflecion Radiomeer (ASER): daa producs for he high spaial resoluion imager on NASA's erra plaform. InJ Remoe Sens 21:847-859
4. Abrams MJ, Brown D (1984) Silver Bell, Arizona, porphyry copper essie repor: ulsa, Oklahoma, he American Associaion of Peroleum Geologiss, he JoinNASA–GeosaesCase Projec. Final Reporchaper 4:4-73
5. Abrams M, Hook SJ (1995) Simulaed ASER daa for geologic sudies. IEEE rans Geosci remoe sens 33(3):
6. Abrams MJ, Brown D, Lepley L, Sadowski R (1983) Remoe sensing of porphyry copper deposis in Souhern Arizona. Econ Geol 78:591-604
7. Abrams M, Hook S, Ramachandran B (2004) ASER User Handbook, Version 2. JePropulsion Laboraory, California Insiue of echnology.
8. Amer R, Kusky , Ghulam A (2010) Lihological mapping in he Cenral Easern Deserof Egypusing ASER daa. J Afr Earh Sci 56:75-82
9. Amer R, Kusky , El Mezayen A (2012) Remoe sensing deecion of gold relaed aleraion zones of Um Rus Area, Cenral Easern Deserof Egyp. Adv Space Res 49:121-134
10. Aminzadeh B, Samani F (2006) Idenifying he boundaries of he hisorical sie of Persepolis using remoe sensing. Remoe Sens Environ 102:52-62

11. Assiri A, Mousa H (2008) Using ASER imagery for massive sulphide deposis exploraion. Microwaves, Radar and Remoe Sensing Symposium, 2008. Proc 2008 IEEE In. 300-303 MRRS
12. Aydal D, Ardal E, Dumanlilar O (2007) Applicaion of he Crosa echnique for aleraion mapping of granioidic rocks using EM + daa: case sudy from easern auride bel(SE urkey). InJ Remoe Sens 28(17):3895-3913
13. Azizi H, arverdi MA, Akbarpour A (2010) Exracion of hydrohermal aleraions from ASER SWIR daa from easZanjan, norhern Iran. Adv Space Res 46:99-109
14. Barry PS, Pearlman J (2001) he EO-1 Misson: Hyperion daa.
15. Barry P, Pearlman JS, Jarecke P, Folkman M (2002) Hyperion daa collecion: performance assessmenand science applicaion. IEEE rans Geosci Remoe Sens 3:1439-1499
16. Beck R (2003) EO-1 User Guide, Version 2.3., Universiy of Cincinnai.
17. Bedini E (2009) Mapping lihology of he Sarfaroq carbonaie complex, souhern WesGreenland, using HyMap imaging specromeer daa. Remoe Sens Environ 113:1208-1219
18. Bedini E (2011) Mineral mapping in he Kap Simpson complex, cenral EasGreenland, using HyMap and ASER remoe sensing daa. Adv Space Res 47:60-73
19. Bedini E, Van Der Meer F, Van Ruienbeek F (2009) Use of HyMap imaging specromeer daa o map mineralogy in he Rodalquilar caldera, souheasSpain. InJ Remoe Sens 30(2):327-348
20. Beroldi L, Massironi M, Visona D, Carosi R, Monomoli C, GulberF, Naleo G, Pelizzo MG (2011) Mapping he Buraburi in he Himalaya of Wesern Nepal: Remoe Sensing analysis in a collisional belwih vegeaion cover and exreme variaion of opography. Remoe Sens Environ 115:1129-1144
21. Bishop CA, Liu JG, Mason PJ (2011) Hyperspecral remoe sensing for mineral exploraion in Pulang, Yunnan Province, China. InJ Remoe Sens 32(9):2409-2426

22. Boardman JW (1989) Inversion of imaging specromery daa using singular value decomposiion. 2069-2072
23. Boardman JW (1992) Sedimenary facies analysis using imaging specromery: a geophysical inverse problem. Unpublished Ph.D. hesis, Universiy of Colorado.
24. Boardman JW (1998) Leveraging he high dimensionaliy of AVIRIS daa for improved sub-pixel argeunmixing and rejecion of false posiives: mixure uned mached filering. Summaries of he Sevenh Annual JPL Airborne Geoscience Workshop, Pasadena, CA. p 55
25. Boardman JW, Kruse FA, Green RO (1995) Mapping argesignaures via parial unmixing of AVIRIS daa, Summaries, Proceedings of he Fifh JPL Airborne Earh Science Workshop. JPL Publicaion, Pasadena, California. pp 23-26 v. 1, p.95 –1
26. Brown J (2000) An adapive sraegy for wavelebased image enhancemen. 67-81
27. BryanR, Moran MS, McElory SA, Holifield C, home KJ, Miura , Biggar SF (2003) Daa Coninuiy of Earh Observing 1(EO-1) Advanced Land Imager (ALI) and LandsaM and EM+. IEEE rans Geosci Remoe Sens 41:6
28. Clark RN (1999) Specroscopy of rocks and minerals, and principles of specroscopy. In: Rencz AN (ed) Remoe sensing for he Earh sciences, Manual of remoe sensing, Volume 3 (hird ediion): New York, John Wiley and Sons, Inc, New York. pp 3-58
29. Clark RN, Roush L (1984) Reflecance specroscopy: quaniaive analysis echniques for remoe sensing applicaions. J Geophys Res 89:6329-6340
30. Clark RN, King VV, Klejwa M, Swayze GA (1990) High specral resoluion reflecance specroscopy of minerals. J Geophys Res 95:12653-12680
31. Clark RN, Swayze GA, Gallagher A, Gorelick N, Kruse 751FA (1991) Mapping wih imaging specromeer daa using he complee band shape leas-squares algorihm simulaneously fio muliple specral feaures from muliple maerials. 2-3

32. Clark RN, Swayze GA, Gallagher A, King VV, Calvin WM (1993) he U.S. Geological Survey, Digial Specral Library: Version 1: 0.2 o 3.0 microns: U.S. Geological Survey Open File Repor. 93-592 1340 p., hp://speclab.cr.usgs.gov *webcie* (Augus1999)

33. Clouis EA (1996) Hyperspecral geological remoe sensing: evaluaion of analyical echniques. InJ Remoe Sens 17(12):2215-2242

34. Cocks , Jenssen R, SewarA, Wilson I, Shields (1998) he HyMap Airborne Hyperspecral Sensor: he Sysem, Calibraion and Performance. In: Schaepman M, Schläpfer D, Ien KI (eds) Proc. 1sEARSeL Workshop on Imaging Specroscopy, EARSeL, Zurich, Paris. pp 37-43

35. Crosa AP, Sabine C, aranik JV (1998) Hydrohermal Aleraion Mapping aBodie, California, using AVIRIS Hyperspecral Daa. Remoe Sens Environ 65:309-319

36. Crosa AP, Souza Filho CR, Azevedo F, Brodie C (2003) argeing key aleraion minerals in epihermal deposis in Paagonia, Argenina, Using ASER imagery and principal componenanalysis. InJ Remoe sens 24:4233-4240

37. Crowley JK, Clark RN (1992) AVIRIS sudy of Deah Valley evaporie deposis using leas- squares band-fiing mehods. Summaries of he hird Annual JPL Airborne Geoscience Workshop, In. pp 29-31

38. Crowley JK, Brickey DW, Rowan LC (1989) Airborne imaging specromeer daa of he Ruby Mounains, Monana: mineral discriminaion using relaive absorpion band-deph images. Remoe Sens Environ 29(2):121-134

39. CudahyJ, Barry PS (2002) Earh magmaic-seawaer hydrohermal aleraion revealed hrough saellie-borne Hyperion imagery aPanorama, vol I. IEEE IGARSS proceeding, Wesern Ausralia. pp 590-592

40. Cudahy J, Hewson R, Huningon JF, Quigley MA, Barry PS (2001) he performance of he saellie-borne Hyperion

hyperspecral VNIR-SWIR imaging sysem for mineral mapping aMounFion. Souh AusIEEE IGARSS proc p. 9-13

41. Di ommaso I, Rubinsein N (2007) Hydrohermal aleraion mapping using ASER daa in he Infiernillo porphyry deposi, Argenina. Ore Geol Rev 32:275-290
42. DucarDF, Crosa AP, Filio CRS (2006) Aleraion mineralogy ahe Cerro La Mina epihermal prospec, Paagonia, Argenina: field mapping, shor-wave infrared specroscopy, and ASER images. Econ Geol 101:981-996
43. El Desouky HA, Muchez P, Dewaele S, Bouwood A, yler R (2008) Posorogenic origin of he sraiform Cu mineralizaion aLufukwe, Lufilian Foreland, Democraic Repulic of Congo. Econ Geol 103:555-582
44. Folkman M, Pearlman J, Liao L, Jarecke P (2001) EO-1/Hyperion hyperspecral imager design, developmen, characerizaion, and calibraion. Hyperspecral Remoe Sensing of he Land and Amosphere. Proc SPIE 4151:40-51
45. Fujisada H (1995) Design and performance of ASER insrumen. In: Breckinridge JB (ed) Proceedings of Inernaional Sociey of Opical Engineering 2583, pp. 16-25
46. Gabr S, Ghulam A, Kusky (2010) Deecing areas of high-poenial gold mineralizaion using ASER daa. Ore Geol Rev 38:59-69
47. Gad S, Kusky (2007) ASER specral raioing for lihological mapping in he Arabian–Nubian shield, he Neoproerozoic Wadi Kid area, Sinai, Egyp. Gondwana Res 11:326-335
48. Galva˜o Le^nio Soares, Raimundo Almeida-Filho, I´caro Viorello (2005) Specral discriminaion of hydrohermally alered maerials using ASER shor-wave infrared bands: Evaluaion in a ropical savannah environmen. InJ Appl Earh Observaion Geoinformaion 7:107-114
49. Gersman R, Ben-Dor E, Beyh M, Avigad D, Abraha M, Kibreba A (2008) Mapping of hydrohermally alered rocks by he EO-1 Hyperrion sensor, norhern Danakil, Erirea. InJ Remoe Sens 29(13):3911-3936

50. Goez AFH (2009) hree decades of hyperspecral remoe sensing of he Earh: A personal view. Remoe Sens Environ 113:S6-S16
51. Goez AFH, Rock BN, Rowan LC (1983) Remoe sensing for exploraion. overview Econ Geol 78:573-590
52. Gomez Ce´cile, Chrisophe Delacour, Pascal Allemand, Parick Ledru, R. Wackerle (2005) Using ASER remoe sensing daa sefor geological mapping, in Namibia. Phys Chem Earh 30:97-108
53. Gonzalez RC, Woods RE (2002) Digial Image Processing. Prenice-Hall Englewood Cliffs, NJ. p 793
54. Goodenough DG, Dyk A, Niemann KO, Pearlman JS, Chen H, Han , Murdoch M, WesC (2003) Processing Hyperion and ALI for ForesClassificaion. IEEE rans Geosci Remoe Sens 41:6
55. Green AA, Berman M, Swizer P, Craig MD (1988) A ransformaion for ordering mulispecral daa in erms of image qualiy wih implicaions for noise removal. IEEE rans Geosci Remoe Sens 26(1):65-74
56. Green RO, Pavri BE, Chrien G (2003) On-OrbiRadiomeric and Specral Calibraion Characerisics of EO-1 Hyperion Derived Wih an Underflighof AVIRIS and *In Siu* Measuremens aSalar de Arizaro, Argenina. IEEE rans Geosci Remoe Sens 41:6
57. Gusafson LB, HunJP (1975) he porphyry copper deposiaEl Salvador, Chile. Econ Geol 70:857-912
58. Harsanyi JC, Farrand WH, Chang CI (1994) Deecion of subpixel signaures in hyperspecral image sequences. Proceedings of 1994 ASPRS Annual Conference, Reno, Nevada. pp 236-247
59. Haselwimmer CE, Riley R, Liu JG (2011) Lihologic mapping in he Oscar II Coasarea, Graham Land, Anarcic Peninsula using ASER daa. InJ Remoe Sens 32(7):2013-2035
60. Hearn DR, Digenis CJ, Lencioni DE, Mendenhall JA, Evans JB, Walesh RD (2001) EO-1 Avanced Land Imager overview and spaial performance. IEEE rans Geosci Remoe Sens p. 897-899
61. Hellman MJ, Ramsey MS (2004) Analysis of hosprings and associaed deposis in Yellowsone Naional Park using ASER

and AVIRIS remoe sensing. J Volcanol Geoherm Res 135:195-219

62. Hewson RD, Cudahy J, Huningon JF (2001) Geological and aleraion mapping aMFion, Souh Ausralia, using ASER saellie-borne daa. IEEE rans Geosci RemoeSens p. 724-726

63. Hewson RD, Cudahy J, Mizuhiko S, Ueda K, Mauger AJ (2005) Seamless geological map generaion using ASER in he Broken Hill–Curnamona province of Ausralia. Remoe Sens Environ 99:159-172

64. Hook SJ, AbboEA, Grove C, Kahle AB, Palluconi F (1999) Use of mulispecral hermal infrared daa in geological sudies. In: Rencz AN (ed) Remoe sensing for he Earh sciences, Manual of remoe sensing, Volume 3 (hird ediion): New York, John Wiley and Sons, Inc, New York. pp 59-110

65. Hosseinjani M, engesani MH (2011) Mapping aleraion minerals using sub-pixel unmixing of ASER daa in he Sarduiyeh area, SE Kerman, Iran. InJ Digial Earh 4(6):487-504

66. Hubbard BE, Crowley JK (2005) Mineral mapping on he Chilean–Bolivian Aliplano using co-orbial ALI, ASER and Hyperion imagery: Daa dimensionaliy issues and soluions. Remoe Sens Environ 99:173-186

67. Hubbard BE, Crowley JK, Zimbelman DR (2003) Comparaive aleraion mineral mapping using visible o shorwave infrared (0.4-2.4 µm) Hyperion, ALI, and ASER imagery. IEEE rans Geosci Remoe Sens 41(6):1401-1410

68. HunG (1977) Specral signaures of pariculae minerals in he visible and near infrared. Geophysics 42:501-513

69. HunGR, Ashley P (1979) Specra of alered rocks in he visible and near infrared. Econ Geol 74:1613-1629

70. HunGR, Salisbury JW (1974) Mid-infrared specral behavior of igneous rocks. echnical ReporAFRCL-R-75-0356, US Air Force Cambridge Research Laboraory, Cambridge, MA.

71. HunGR, Salisbury JW, Lenhoff CJ (1971a) Visible and near-infrared specra of minerals and rocks: III. Oxides hydroxides Modern Geol v 2:195-205

72. HunGR, Salisbury JW, Lenhoff CJ (1971b) Visible and near-infrared specra of minerals and rocks: IV. Sulphides sulphaes Modern Geol v 3:1-14

73. Huningon JF (1996) he role of remoe sensing in finding hydrohermal mineral deposis on Earh. Evoluion of Hydrohermal Ecosysems on Earh (and Mars?. Wiley, England. pp 214-234

74. Inzana J, Kusky , Higgs G, ucker R (2003) Supervised classificaions of LandsaM band raio images and LandsaM band raio image wih radar for geological inerpreaions of cenral Madagascar. J Afr Earh Sci 37:59-72

75. James K, Vergo C, Vergo N (1988) Near-infrared reflecance specra of mixures of kaolin-group minerals: use in clay mineral sudies. Clay and Clay Miner 36(4):310-316

76. Junek P (2004) Geological mapping in he Cheleken Peninsula, urkminisan area using Advanced Spaceborne hermal Emission and Reflecion Radiomeer (ASER) daa. ISPRS Conference, California, USA.

77. Kargi H (2007) Principal componens analysis for borae mapping. InJ Remoe Sens 28(8):1805-1817

78. Kavak KS (2005) Recogniion of gypsum geohorizons in he Sivas Basin (urkey) using ASER and LandsaEM + images. InJ Remoe Sens 26(20):4583-4596

79. Khan SD, Mahmood K, Casey JF (2007) Mapping of Muslim Bagh ophiolie complex (Pakisan) using new remoe sensing, and field daa. J Asian Earh Sci 30:333-343

80. KraC, Calvin WM, Coolbaugh MF (2010) Mineral mapping in he Pyramid Lake basin: Hydrohermal aleraion, chemical precipiaes and geohermal energy poenial. Remoe Sens Environ. doi:10.1016/j.rse.2010.05.006

81. Kruse FA (2003) Mineral Mapping wih AVIRIS and EO-1 Hyperion. Presened ahe 12h JPL Airborne Geoscience Workshop, Pasadena, California.

82. Kruse JW, Boardman FA (2000) Characerizaion and Mapping of Kimberlies and Relaed Diaremes Using Hyperspecral Remoe

Sensing. Proceedings, 2000 IEEE Aerospace Conference, Big Sky, MO. pp 18-24

83. Kruse FA, Perry SL (2007) Regional mineral mapping by exending hyperspecral signaures using mulispecral daa. IEEE rans Geosci Remoe Sensi v 4:1-14

84. Kruse FA, Boardman JW, Lefkoff AB, HeidebrechKB, Shapiro A, Barloon PJ, Goez AFH (1993) he Specral Image Processing Sysem (SIPS) – Ineracive Visualizaion and Analysis of Imaging Specromeer Daa. Remoe Sens Environ 44:145-163

85. Kruse FA, Boardman JW, Huningon JF (1999) Fifeen Years of Hyperspecral Daa: Norhern Grapevine Mounains, Nevada: in Proceedings of he 8h JPL Airborne Earh Science Workshop: JePropulsion Laboraory Publicaion. JPL Publ 99–17:247-258

86. Kruse FA, Perry SL, Caballero A (2006) Disric-level mineral survey using airborne hyperspecral daa, Los Menucos, Argenina. Ann Geophys 49(1):83-92

87. Kucukkaya AG (2004) Phoogrammery and remoe in archeology. J QuanSpecroscopy Radiaive ransf 88:83-88

88. Kusky M, Ramadan M (2002) Srucural conrols on Neoproerozoic mineralizaion in he Souh Easern Deser, Egyp: an inegraed field, LandsaM, and SIR-C/X SAR approach. J Afr Earh Sci 35:107-121

89. Lammoglia , Filho CRS (2011) Specroscopic characerizaion of oils yielded from Brazilian offshore basins: Poenial applicaions of remoe sensing. Remoe Sens Environ 115:2525-2535

90. Lencioni DE, Digenis CJ, Bicknell WE, Hearn DR, Mendenhall JA (1999) Design and performance of he EO1 Advanced Land Imager. SPIE conference on sensors, sysems, and nexgeneraion saellies III, Florence, Ial.

91. Liao L, Jarecke P, Gleichauf D, Hedman (2000) Performance characerizaion of he Hyperion imaging specromeer insrumen. Proc SPIE 4135:264-275

92. Lobell DB, Asner GP (2003) Comparison of Earh Observing-1 ALI and LandsaEM + for Crop Idenificaion and Yield Predicion in Mexico. IEEE rans Geosci RemoeSens 41:6

93. Lowell JD, GuilberJM (1970) Laeral and verical aleraion-mineralizaion zoning in porphyry ore deposis. Econ Geol Bull Soc Econ Geol 65(4):373-408

94. Mars JC, Rowan LC (2006) Regional mapping of phyllic- and argillic-alered rocks in he Zagros magmaic arc, Iran, using Advanced Spaceborne hermal Emission and Reflecion Radiomeer (ASER) daa and logical operaor algorihms. Geosphere 2(3):161-186

95. Mars JC, Rowan LC (2010) Specral assessmenof new ASER SWIR surface reflecance daa producs for specroscopic mapping of rocks and minerals. Remoe Sens Environ 114:2011-2025

96. Mars JC, Rowan LC (2011) ASER specral analysis and lihologic mapping of he Khanneshin carbonae volcano, Afghanisan. Geosphere 7:276-289

97. Massironi ML, Beroldi P, Calafa D, Visona A, Bisacchi C, Giardino A, Schiavo B (2008) Inerpreaion and processing of ASER daa for geological mapping and granioids deecion in he Saghro massif (easern Ani-Alas, . Morocco Geosphere 4(4):736-759

98. Mendenhall JA, Ryanhoward DP, Willard BC (2000) Earh Observing-1 Advanced Land Imager: insrumenand flighoperaions overview. MILincoln Lab ProjecReporEO1–1, 23.

99. Moghaderi A, Moore F, Mohammadzadeh A (2007) he applicaion of advanced space-borne hermal emission and reflecion (ASER) radiomeer daa in he deecion of aleraion in he Chadormalu paleocraer, Bafq region, Cenral Iran. J Asian Earh Sci 30:238-252

100. Moore F, Rasmanesh F, Asady H, Modabberi S (2008) Mapping mineralogical aleraion using principal componenanalysis and mached filer processing in akab area, norh-weslran, from ASER daa. InJ Remoe Sens 29(10):2851-2867

101. Naional Aeronauics and Space Adminisraion (2002) Earh Observing-1 Advanced Land Imager.

102. Naional Aeronauics and Space Adminisraion (2004) Earh Observing-1 EO1General Mission. Online: hp://eo1.gsfc. nasa.gov/new/general/ *webcie*

103. Ninomiya Y, Fu B, Cudahy J (2005) Deecing lihology wih Advanced Spaceborne hermal Emission and Reflecion Radiomeer (ASER) mulispecral hermal infrared "radiance-a-sensor" daa. Remoe Sens Environ 99(1–2):127-139

104. Okada K, Segawa K, Hayashi I (1993) Removal of he vegeaion effecfrom LANDSAM and GER imaging specroradiomeer daa. ISPRS J Phoogramm Remoe Sens 48(6):16-27

105. Ozan NS, Suzen ML (2011) Mapping evaporae minerals by ASER. InJ Remoe Sens 32(6):1651-1673

106. Pearlman JS, Barry PS, Segal CC, Shepanski J, Beiso D, Carman SL (2003) Hyperion, a Space-Based Imaging Specromeer. IEEE rans Geosci Remoe Sens 41(6):1160-1173

107. Perry SL (2004) Spaceborne and airborne remoe sensing sysems for mineral exploraion-case hisories using infrared specroscopy. In: King PL, Ramsey MS, Swayze GA (eds) Infrared Specroscopy in Geochemisry, Exploraion Geochemisry, and Remoe Sensing, Mineralogic Associaion of Canada, London, Canada. pp 227-240

108. Perovic A, Khan SD, hurmond AK (2012) Inegraed hyperspecral remoe sensing, geochemical and isoopic sudies for undersanding hydrocarbon-induced rock aleraions. Mar PeGeol 35(2012):292-308

109. Pieri D, Abrams M (2004) ASER waches he world's volcanoes: a new paradigm for volcanological observaions from orbi. J Volcanol Geoherm Res 135:13-28

110. Podwysocki MH, Mimms DL, Salisbury JW, Bender LV, Jones OD (1984) Analysis of Landsa-4 M daa for lihologic and image mapping purpose, Proceedings of Landsa-4 Science Invesigaions Summary. Greenbel, Maryland 2:35-39

111. Pour BA, Hashim M (2011a) Idenificaion of hydrohermal aleraion minerals for exploring of porphyry copper deposiusing ASER daa, SE Iran. J Asian Earh Sci 42:1309-1323

112. Pour BA, Hashim M (2011b) Specral ransformaion of ASER and he discriminaion of hydrohermal aleraion minerals in a semi-arid region, SE Iran. InJ Physical Scie 6(8):2037-2059
113. Pour BA, Hashim M (2011c) Applicaion of Spaceborne hermal Emission and Reflecion Radiomeer (ASER) daa in geological mapping. InJ Physical Sci 6(33):7657-7668
114. Pour BA, Hashim M (2011d) he Earh Observing-1 (EO-1) saellie daa for geological mapping, souheasern segmenof he Cenral Iranian Volcanic Bel, Iran. InJ Physical Scie 6(33):7638-7650
115. Pour BA, Hashim M (2012a) he applicaion of ASER remoe sensing daa o porphyry copper and epihermal gold deposis. Ore Geol Rev 44:1-9
116. Pour BA, Hashim M (2012b) Idenifying areas of high economic-poenial copper mineralizaion using ASER daa in Urumieh-Dokhar Volcanic Bel, Iran. Adv Space Res 49:753-769
117. Pour BA, Hashim M, Marghany M (2011) Using specral mapping echniques on shorwave infrared bands of ASER remoe sensing daa for aleraion mineral mapping in SE Iran. InJ Physical Sci 6(4):917-929
118. Qiu F, Abdelsalam M, hakkar P (2006) Specral analysis of ASER daa covering parof he Neoproerozoic Allaqi-Heiani suure, Souhern Egyp. J Afr Earh Sci 44:169-180
119. Rajendran S, hirunavukkarasu A, Balamurugan G, Shankar K (2011) Discriminaion of iron ore deposis of granulie errain of Souhern Peninsular India using ASER daa. J Asian Earh Sci 41:99-106
120. Rajendran S, Al-Khirbasha S, Pracejusa B, Nasira S, Al-Abria AH, Kusky M, Ghulam A (2012) ASER deecion of chromie bearing mineralized zones in Semail Ophiolie Massifs of he norhern Oman Mounains: Exploraion sraegy. Ore Geol Rev 44:121-135

121. Rajesh HM (2008) Mapping Proerozoic unconformiy-relaed uranium deposis in he Rockole area, Norhern erriory, Ausralia using LandsaEM+. Ore Geol Rev 33:382-396

122. Ramadan M, Abdel Faah MF (2010) Characerizaion of gold mineralizaion in Garin Hawal area, Kebbi Sae, NW Nigeria, using remoe sensing. EgypJ Remoe Sens Space Sci 13:153-163

123. Ren D, Abdelsalam MG (2006) racing along-srike srucural coninuiy in he Neoproerozoic Allaqi-Heiani Suure, souhern Egypusing principal componenanalysis (PCA), fasFourier ransform (FF), and redundanwaveleransform (RW) of ASER daa. J Afr Earh Sci 44:181-195

124. Rockwell BW, Hofsra AH (2008) Idenificaion of quarz and carbonae minerals across norhern Nevada using ASER hermal infrared emissiviy daa Implicaions for geologic mapping and mineral resource invesigaions in well-sudied and fronier areas. Geosphere 4(1):218-246

125. Rowan LC, Mars JC (2003) Lihologic mapping in he Mounain Pass, California area using Advanced Spaceborne hermal Emission and Reflecion Radiomeer (ASER) daa. Remoe Sens Environ 84:350-366

126. Rowan LC, Welaufer PH (1981) Relaion beween regional lineamensysems and srucural zones in Nevada. Am Assoc PeGeol Bull 65:1414-1432

127. Rowan LC, Welaufer PH, Goez AFH, Billingsley FC, SewarJH (1974) Discriminaion of rock ypes and deecion of hydrohermally alered areas in souh-cenral Nevada. US Geol Survey Prof Paper 883:35

128. Rowan LC, Goez AFH, Ashley RP (1977) Discriminaion of hydrohermally alered and unalered rocks in visible and near infrared mulispecral images. Geophysics 42(3):522-535

129. Rowan LC, Hook SJ, Abrams MJ, Mars JC (2003) Mapping hydrohermally alered rocks aCuprie, Nevada, using he Advanced Spaceborne hermal Emission and Reflecion

Radiomeer (ASER), a new saellie-imaging sysem. Econ Geol 98(5):1019-1027

130. Rowan LC, Mars JC, Simpson CJ (2005) Lihologic mapping of he Mordor N., Ansralia ulramafic complex by using he Advanced Spaceborne hermal Emission and Reflecion Radiomeer (ASER). Remoe sensing of Environmen99:105-126

131. RowanLC, RoberGS, JohnC(2006) Disribuionofhydrohermally alered rocks in he Reko Diq, Pakisan mineralized area based on specral analysis of ASER daa. Remoe Sens Environ 104:74-87

132. Sabins FF (1996) Remoe Sensing Principles and Inerpreaion. Freeman & Co, New York, USA.

133. Sabins FF (1997) Remoe sensing sraegies for mineral exploraion. In: Rencz AE (ed) Remoe Sensing for he Earh Sciences, John Wiley & Sons, Inc., New York. pp 375-447

134. Sabins FF (1999) Remoe sensing for mineral exploraion. Ore Geol Rev 14:157-183

135. Salai S, Van Ruienbeek FJA, Van der Meer FD, angesani MH, Van der Werff H (2011) Lihological mapping and fuzzy seheory: Auomaed exracion of lihological boundary from ASER imagery by emplae maching and spaial accuracy assessmen. Inernaional Journal of Applied Earh Observaion and Geoinformaion 13:753-765

136. Salisbury JW, D'Aria DM (1992a) Emissiviy of erresrial maerials in he 8–14 μm amospheric window. Remoe Sens Environ 42:83-106

137. Salisbury JW, D'Aria DM (1992b) Infrared (8–14 μm) remoe sensing of soil paricle size. Remoe Sens Environ 42:157-165

138. Sanjeevi S (2008) argeing limesone and bauxie deposis in Souhern India by specral unmixing of hyperspecral image daa. he Inernaional Archives of he Phoogrammery, vol XXXVII. Remoe Sensing and Spaial Informaion Sciences, ParB8. Beijing.

139. Shi P, Fu B, Ninomiya Y, Sun J, Li Y (2012) Mulispecral remoe sensing mapping for hydrocarbon seepage-induced lihologic

anomalies in he Kuqa foreland basin, souh ian Shan. J Asian Earh Sci 46:70-77

140. Sillioe RH (2010) Porphyry Copper Sysems. Economic Geology 105:3-41
141. Singh A, Harrison A (1985) Sandardized principal componens. InJ Remoe Sens 6:883-896
142. Smailbegovic A, aranik JV (1999) Dollars and sensing – he Economics of Remoe Sensing in Economic Geology Applicaion. In: Rogres RH (ed) Proceedings of he hireenh Inernaional Conference on Applied Geologic Remoe Sensing, Vancouver, Briish Columbia,Canada. pp 1-3195–203
143. Spaz DM, Wilson R(1995) Remoe sensing characerisics of porphyry copper sysems, wesern America Cordillera. In: Pierce FW, Bolm JG (eds) Arizona Geological Sociey Diges, vol 20, pp. 94-108
144. Sulan M, Arvidson RE, Surchio NC, Guinness EA (1987) Lihologic mapping in arid regions wih Landsahemaic mapper daa: Meaiq Dome, Egyp. Geol Soc Am Bull 99(6):748-762
145. angesani MH, Moore F (2000) Iron oxide and hydroxyl enhancemenusing he Crosa Mehod: a case sudy from he Zagros Bel. Fars province, Iran, Inernaional Journal of Applied Earh Observaion and Geoinformaion 2:140-146
146. angesani MH, Mazhari N, Ager B, Moore F (2008) Evaluaing advance spaceborne hermal emission and reflecion radiomeer (ASER) daa for aleraion zone enhancemenin a semi-arid area, norhern shahr-e-Babak, SE Iran. InJ Remoe Sens 29(10):2833-2850
147. angesani MH, Jaffari L, VincenRK, Sridhar BBM (2011) Specral characerizaion and ASER-based lihological mapping of an ophiolie complex: A case sudy from Neyriz ophiolie, SW Iran. Remoe Sens Environ 115:2243-2254
148. home KJ, Biggar SF, Wisniewski W (2003) Cross Comparison of EO-1 Sensors and Oher Earh Resources Sensors o Landsa-7 EM+Using Railroad Valley Playa. IEEE rans Geosci Remoe Sens 41:6

149. iley SR (1972) Inrusion and wallrock porphyry copper deposis. Econ Geol Bull Soc Econ Geol 67:122

150. Ungar SG, Pearlman JS, Mendenhall JA, Reuer D (2003) Overview of he Earh Observing One (EO-1) Mission. IEEE rans Geosci Remoe Sens 41(6):1149-1159

151. Van der Meer FD, Van der Werff HMA, Van Ruienbeek FJA, Hecker CA, Bakker WH, Noomen MF, Van der Meijde M, Carranza EJM, Boudewijn de Smeh J, Woldai (2012) Muli- and hyperspecral geologic remoe sensing: A review. Inernaional Journal of Applied Earh Observaion and Geoinformaion 2012:112-128

152. Van der Werff H, van Ruienbeek F, van der Meijde M, van der Meer F, de Jong S, Kalubandara S (2007) Roaion-varianemplae maching for supervised hyperspecral boundary deecion. Geosci Remoe Sens Leers IEEE 4:70-74

153. Van Ruienbeek FJA, Cudahy J, Van der Meer FD, Hale M (2012) Characerizaion of he hydrohermal sysems associaed wih Archean VMS - mineralizaion aPanorama, Wesern Ausralia, using hyperspecral, geochemical and geohermomeric daa. Ore geology reviews 45:33-46

154. Vapnik VN (1995) he Naure of Saisical Learning heory. Springer-Verlag, New York.

155. Vaughan RG, Hook SJ, Calvin WM, aranik JV (2005) Surface mineral mapping aSeamboaSprings, Nevada, USA, wih muli-wavelengh hermal infrared images. Remoe Sens Environ 99:140-158

156. Velosky JC, Sern RJ, Johnson PR (2003) Geological conrol of massive sulfide mineralizaion in he Neoproerozoic Wadi Bidah shear zone, souhwesern Saudi Arabia, inferences from orbial remoe sensing and field sudies. Precambrian Res 123(2–4):235-247

157. Vicene LE, Filho CRS (2011) Idenificaion of mineral componens in ropical soils using reflecance specroscopy and advanced spaceborne hermal emission and reflecion radiomeer (ASER) daa. Remoe Sens Environ 115:1824-1836

158. Waldhoff G, Bubenzer O, Bolen A, Koppe W, Bareh G (2008) Specral analysis of ASER, Hyperion, and Quickbird daa for geomorphological and geological research in Egyp(Dakhla Oasis, Wesern deser). In: he Inernaional Archives of Phoogrammery, Remoe Sensing and Spaial Informaion Sciences, vol XXXVII. ParB8. Beijing. pp 1201-1206

159. Was DR, Harris NBW (2005) Mapping granie and gneiss in domes along he Norh Himalayan aniform wih ASER SWIR band raios. Geol Soc Am Bull 117(7/8):879-886

160. Wulder MA, Whie JC, Goward SN, Jeffrey GM, Irons JR, Herold M, Cohen WB, Loveland R, Woodcock CE (2008) Landsaconinuiy: Issues and opporuniies for land cover monioring. Remoe Sens Environ 112:955-969

161. Xu Y, Qizhong L, Shao Y, Wang L (2004) Exracion Mechanism of Aleraion Zones using ASER Imagery. Geosciences and Remoe Sensing Symposium, IGARSS 04, Proceedings 2004 IEEE Inernaional, vol 6. 4174-4175

162. Yamaguchi YI, Fujisada H, Kudoh M, Kawakami , su H, Kahle AB, Pniel M (1999) ASER insrumencharacerizaion and operaion scnorio. Adv Space Res 23(8):1415-1424

163. Yamaguchi YI, Fujisada H, Kahle AB, su H, Kao M, Waanabe H, Sao I, Kudoh M (2001) ASER insrumenperformance, operaion saus, and applicaion o Earh sciences. IEEE rans Geosci Remoe Sensi p. 1215-1216

164. Yu L, Porwal A, Holden EJ, Denih MC (2012) owards auomaic lihological classificaion from remoe sensing daa using supporvecor machines. CompuGeosci 45:229-239

165. Yujun Z, Jianmin Y, Fojun Y (2007) he Poenials of Muli-specral Remoe Sensing echniques for Mineral Prognosicaion – aking Mongolian Oyu olgoi Cu-Au Deposias an Example. Earh Sci Froniers 14(5):63-70

166. Zhang X, Pazner M (2007) Comparison of Lihologic Mapping wih ASER, Hyperion, and EM Daa in he Souheasern Chocolae Mounains, USA. Phoogramm Eng Remoe Sens 73(5):555-561

167. Zhang X, Panzer M, Duke N (2007) Lihologic and mineral informaion exracion for gold exploraion using ASER daa in he souh Chocolae Mounains (California). J Phoogrammery and Remoe Sens 62:271-282

168. Zoheir B, Emam A (2012) Inegraing geologic and saellie imagery daa for high-resoluion mapping and gold exploraion arges in he Souh Easern Deser, Egyp. J Afr Earh Sci 66–67:22-34

Chapter 7

Petroleum Geology of the Puguang Sour Gas Field in the Sichuan Basin, Sw China

Yongsheng Ma[a], Shuichang Zhang[b], Tonglou Guo[a], Guangyou Zhu[a], Xunyu Cai[a], Maowen Li[c]

[a]SINOPEC Southern Exploration and Development Company, Kunming, Yunnan 650200, China

[b]PetroChina Research Institute of Petroleum Exploration and Development, Beijing 100083, China

[c]Geological Survey of Canada, 3303-33 Street NW Calgary, Alta., Canada T2L 2A7

ABSTRACT

The Puguang sour gas field in northeastern Sichuan Basin is the largest gas discovery in the Paleozoic marine strata in China, with at least 9 trillion cubic feet (tcf) of proven gas reserves. It is one of the many anticlinal structures formed during the Indo-Sinian orogeny, with gas being trapped in a composite structure–lithology pool. The presence of the Upper Permian coal measures, Lower Silurian and Lower Cambrian marine shales provided excellent hydrocarbon source rocks. The Lower Triassic and Upper Permian oolitic dolomites serve as high quality reservoir rocks, with 10–25% porosity and 1–1000 mD permeability, largely due to diagenetically produced secondary porosity. The Lower and Middle Triassic evaporate beds have acted as the direct top seal for the large Lower Triassic–Upper Permian gas accumulation. The thick Upper Triassic and Jurassic mudstone beds regionally cap the underlying marine carbonate strata. The deep-seated basement faults acted as vertical hydrocarbon migration conduits during faulting, and became effective flow barriers when the faulting ceased. The deep burial of a paleo-oil accumulation in reservoirs rich in sulfate resulted in extensive thermochemical sulfate reduction (TSR). This, together with the new gas injection and structural reconfiguration, transformed a paleo-oil pool to a large sour gas pool.

INTRODUCTION

Many of the Paleozoic marine carbonate strata in southern China have experienced multicyclic tectonic events, and are often characterized by over-matured source rock, poor reservoir quality, high structural complexity, and less ideal preservation conditions. As a result, oil and gas distributions in southern China are complex, presenting great challenges for petroleum exploration (Li et al., 1998; Zhang et al., 2002; Dai et al., 2005; Jin, 2005). It has been known for years that excellent petroleum source rocks are present in a number of basins developed on the ancient Yangtze

Paraplatform (Huang, 1954). However, only small to medium oil and gas discoveries were made in many of the basins.

One notable exception is the Sichuan Basin in southwestern China. Following over 50 years' active petroleum exploration, around 300 gas pools have been discovered there (Zhu et al., 2006a), with a cumulative proved gas reserve around 1 trillion cubic meters (tcm) or 35 tcf by the end of 2005 (Wang et al., 2005). The exploration efforts were often challenged by the difficulty in finding high quality reservoirs. Recent drillings into the Lower Triassic Feixianguan Formation in northeastern Sichuan Basin have led to the initial discovery of Luojiazhai, Dukouhe and Tieshanpo gas pools, followed by the confirmation of the giant Puguang gas field (Ran et al., 2005; Zhang et al., 2006). These encouraging results reveal the presence of high-quality carbonate reservoirs in the deep marine strata of this basin and significantly larger petroleum potential than previously anticipated.

The Puguang gas field was discovered from one of the many structures located in the Shuangshimiao–Puguang structure belt of the Xuhan–Daxian region, northeastern Sichuan Basin (Fig. 1). By the end of 2005, the proven gas reserves in the Puguang field had reached 251 billion cubic meters (bcm) or 8.87 trillion cubic feet (tcf), of which almost 87% was in the Feixianguan Formation and the rest in the Changxing Formation. The gas payzones are dominantly in the thick-bedded reef and dolomitized carbonate shoal reservoirs, with over 400 m of effective gas column. The maximum single well daily production obtained is up to 1.28 million cubic meters (mcm) or 45.2 million cubic feet (mcf) (Ma et al., 2005a and Ma et al., 2005b). More drilling is currently on the way to broaden the gas-bearing area and delineate the ultimate gas reserves.

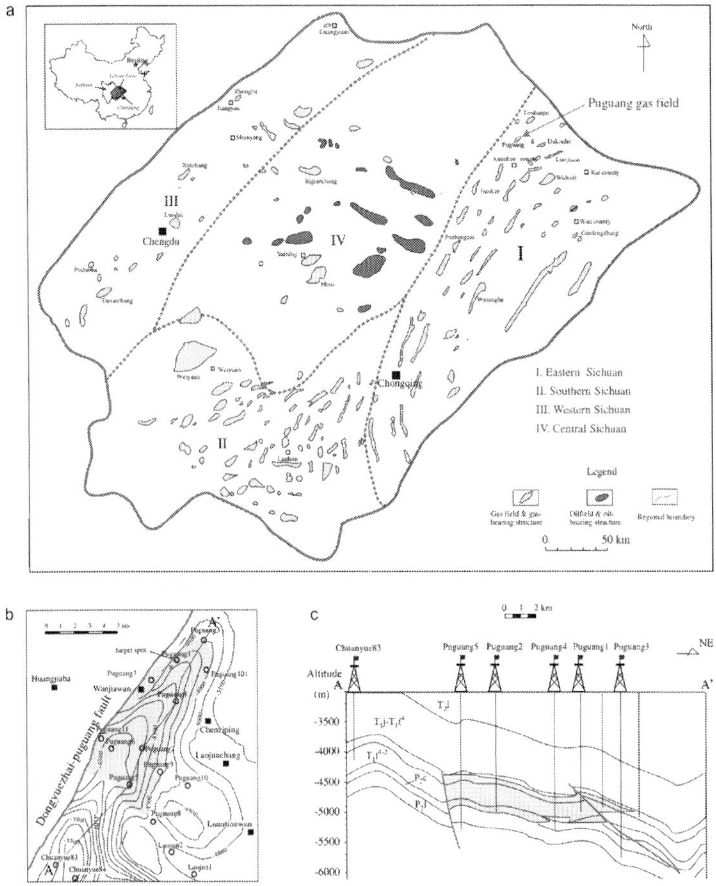

Figure 1: (a) Oil and gas fields in the Sichuan Basin; (b) structural contour map of the Puguang gas field, and (c) a structural cross section along the Puguang gas field.

The objective of this manuscript is to describe the basic elements of the petroleum systems in the Puguang gas field, discuss prevailing geological factors controlling the formation of this field, and on the basis of regional geological setting propose potential areas for future petroleum exploration in the northeastern Sichuan Basin.

GEOLOGICAL SETTING

Basin Evolution

The Sichuan Basin, with an area of approximately 190,000 km^2, is a large cratonic basin, containing up to 6–12 km of Sinian to Quaternary sediments (Fig. 2; Zhai, 1992; Li et al., 2005). The basement of the basin, usually referred to as the Yangtze Paraplatform (Huang, 1954) is a united craton, which was formed during the Jinning (850 Ma) and Chengjiang (700 Ma) orogenies and then crystallized and metamorphosed before the Late Sinian (Wang et al., 2002).

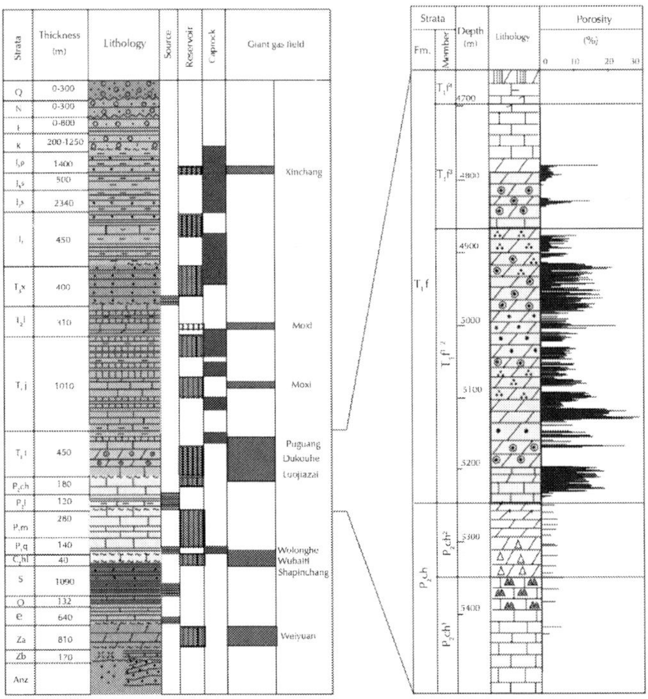

Figure 2: Stratigraphic column of the Sichuan Basin, showing the key petroleum system elements for the Upper Permian–Lower Triassic gas pools.

The tectonic evolution of the Sichuan Basin has experienced a transition from a relatively stable craton to a less stable craton stage (Wang, 1989). As a result, regional elevations and subsidence resulted in large-sized uplift and depressional structures during the Early Paleozoic (up to the Early Permian). Beginning from the Devonian, tensional fault depression first took place along the cratonic margins on the background of the Caledonian uplifts and depressions and gradually extended into the platform. Basic eruptions occurred along certain rift-depression zones through to the Middle Carboniferous, and the pull-apart movement reached its climax in the Permian with large-scale basalt eruption over much of the Upper Yangtze Craton. In the Early and Middle Triassic, the cratonic margins were raised because of the compression from the Tethys and the Pacific, producing algal banks and oolitic shoals that blocked the seawater and turned the central part of the basin into a saline basin. At the end of Triassic, several episodes of the Indo-Sinian orogeny resulted in large folds, and the continental sedimentary development stage began in Sichuan. From the Jurassic to the Early Cretaceous, the western basin margins were uplifted several times, and each uplift was accompanied by strong depression in the mountain front. This led to a continental basin configuration that was characterized by a uplifted center surrounded by a depression. Between the early and Late Cretaceous, the eastern Sichuan region was uplifted, with no sediment being deposited ever since. In contrast, continuous sediments were deposited during the Cretaceous and the Neogene in the western Sichan, with the lake basin gradually shrinking back to the west. The Himalayan orogeny resulted in three strong episodes of folding in western Sichuan. This orogeny corresponds to the closing period of the Tethys and the influence of Pacific Plate subduction shows itself clearly in the eastern Sichuan. The counter-clockwise coupling was the fundamental stress causing the formation of the Sichuan rhombic tectonic basin.

Stratigraphy

The stratigraphy of the Sichuan Basin was summarized in detail by Wang et al. (1989). Due to the successive tectonic movements, some of the sedimentary successions in the study area were thinned or eroded away. In general, the Paleozoic strata have been well preserved, with the exception of the Middle Cambrian to Lower Ordovician strata in the Micang Mountains and the Middle Silurian to Carboniferous strata in most of the northeastern Sichuan Basin, as the result of the Caledonian orogeny. The total thickness of the Lower Palaeozoic strata is about 2000–4000 m in the study area, while the Upper Palaeozoic strata are only about 400–1000 m in thickness.

The Lower Sinian strata consist largely of arkose and shales, whereas the upper Sinian strata include mainly dolomite and shale interbeds. In contrast, the Lower Cambrian strata are composed of interbedded black and gray green shales, carbonate and siliceous rocks deposited in open marine-continental shelf settings. The Middle and Upper Ordovician sediments are dominantly thick limestones, dark carbonaceous shales and silicates. The lower part of the Silurian Longmaxi Formation consists of black shales rich in graptolite that were deposited in open marine continental shelf facies. The remnants in the Middle Silurian strata are composed mainly of gray and yellow shales, sandy shales and mauve shales. These were followed by a large hiatus in much of the Devonian, with the Carboniferous Huanglong Formation being developed only in parts of the eastern basin (Hu and Xie, 1997; Liu et al., 1997). The sediments in the Huanglong Formation range regionally from 0 to 90 m, and 10 to 20 m in the Puguang field area, composed mainly of bioclastic limestones and dolomite breccias.

Following a long period of erosion, seawater transgressed from the southwest in the Early Permian, submerging the entire platform, and carbonates of open neritic platform facies were deposited all over the Sichuan Basin to a thickness of several hundred meters. By the Late Permian, large scale faulting had led to extensive basalt eruption and diabase intrusion in the eastern and western parts of the basin, with continental facies and littoral swamp facies being developed in the southwestern basin and carbonate platform facies in most parts of the basin.

In the Early and Middle Triassic, the surrounding areas of the carbonate platform were relatively rising, leading to a restricted sea and many polyphase evaporate-bearing basins. The evaporate are mainly microcrystalline and micritic, and bioclastics and other grained rocks are present only in small proportions. In the surrounding areas of the rhombic block, there were extremely shallow-water algal banks and oolitic shoals deposited, and these became the significant petroleum reservoir rocks. Thus the process from the vast transgression in the Early Permian to the salting and shrinkage of the sea basin in the Middle Triassic formed a complete and sequential sedimentary cycle, with the changes in paleography and sedimentary environment occurring much more frequently than in the Early to Middle Paleozoic. The linear distribution of the facies belt showed significant faulting control on the craton.

The Upper Triassic strata were deposited mainly in transitional facies, consisting of thick-bedded sandstone, argillutite, siltstone and coal interbeds in the Xujiahe Formation. The significant change in depositional environment of the Upper Triassic were mainly observed in the western basin margins, whilst the central and eastern portions of the basin has always been relatively stable with alternating deposits of fluvial and lacustrine-bog facies. After the late episode of the Indo-Sinian orogeny, the northern and western margins of the continental basin were basically shaped. Since then, many sets of piedmont alluvial-fan conglomerate were accumulated due to the periodical rises. The rise of the old land occurred also in northeastern and southwestern basin margins. Only the southern basin continued to be a lake basin, and the sediments were largely purplish-red mudstone intercalated with river-channel sandstone.

The Middle and Lower Jurassic encountered widely in outcrops in southeastern Sichuan are composed of mauve mudstone and gray argillites interbedded thin coal beds. At the end of the Early Cretaceous, the eastern basin was extensively folded and then uplifted, while the sedimentary basin shrank back to a corner in west Sichuan with the Neogene red fluvial and lacustrine sediments being deposited on top of the Upper Cretaceous strata. After the Himalayan orogeny, the western Sichuan developed into a polyphase depression basin, with the maximum thickness of the Quaternary sediments reaches 300 m.

Oil and Gas Distribution

Based on the basement structure, sedimentary covers, and petroleum occurrence, the Sichuan Basin can be divided into four oil and gas accumulation districts (i.e. the Eastern, Southern, Western and Central oil districts), with nearly 70% of the discovered gas reserves in the first two districts (Wang et al., 1989). Due to the shift in depocenters and the heterogeneity in basin evolution, different gas districts exhibit rather distinct sedimentary sequences and thus different oil and gas distributions. Gas pools occur in every district, but oil is seen only in the Central district. The gas pools in the Southern district are generally small in size, whilst the Eastern district contains most of the known large and medium gas fields (Wang et al., 2005).

PETROLEUM SYSTEM ELEMENTS OF THE PUGUANG GAS FIELD

The Puguang gas field is located in the northern segment of the Eastern gas district (Fig. 1). It is a large structure–lithology composite gas pool with a confirmed gas enclosure area of approximately 50 km^2. The main gas payzones include dolomite reservoirs in the Lower Triassic Feixianguan Formation and Upper Permian Changxing Formation, developed along the platform edge (Zhao et al., 2005; Wang et al., 2002;Ma et al., 2005a, Ma et al., 2005b and Ma et al., 2006). The Oolitic dolomite in the Feixianguan Formation and biohermal limestone in the Changxing Formation overlap laterally and connect to each other vertically, with wide aerial distribution. The present burial depth of the gas reservoirs in the Feixianguan

Formation are relatively deep, averaged at 5200 m. All of the six wells that were drilled in the Puguang gas field produced high flows, approximately 0.5–1.2 mcm/d before acidification, confirming the presence of thick high-quality gas reservoirs.

Production tests of the Puguang 1 and 2 wells show that gas reservoirs in the Feixianguan Formation belong to a single pressure system. The reservoirs in the Changxing Formation and the third member of the Feixianguan Formation are in good communication, as shown by a narrow range in pressure coefficient (1.0–1.12). According to the measured temperature values from the Puguang 1 and 2 wells, the calculated geothermal gradient of the gas reservoir is around 2.18/100 m, consistent with the regional average (2/100 m) (Wang et al., 1989). Thus, the Feixianguang Formation reservoirs in the Puguang gas field currently possess an abnormally low geothermal gradient. Natural gases produced from the Puguang field are typically dry, with little ethane and virtually no propane and higher alkanes, being one of the driest gases produced in any Chinese gas field. The concentrations of H_2S in gases produced from the Puguang field range from 14% to 17% among the highest in China (Zhu et al., 2004). The porosity values of the carbonate reservoirs with burial depth around 5000 m are up to 20%. This and over 100 m of gas column make them carbonate reservoirs one of the deepest reservoirs with the best secondary porosity in China.

Hydrocarbon Source Rocks

Several potential petroleum source rocks occur in northeastern Sichuan Basin, including the Lower Cambrian Jiulaodong Formation, the Lower Silurian Longmaxi Formation and the Upper Permian Longtan Formation (Dai, 1985; Dai et al., 2000 and Dai et al., 2001; Zhu et al., 2006a). The Jiulongdong Formation contains about 200 m of gray to black shales and mudstones that are relatively rich in algal organic matter (with average TOC around 1.67%). These rocks are currently overmature with respect to liquid hydrocarbon generation (with ~3.5% Ro), acting mainly as a gas source rock. The Longmaxi Formation consists of graptolite-bearing black shales and mudstones with high TOC contents (Fig. 3), and is considered to be the main source rock for the Carboniferous reservoired gases in eastern Sichuan.

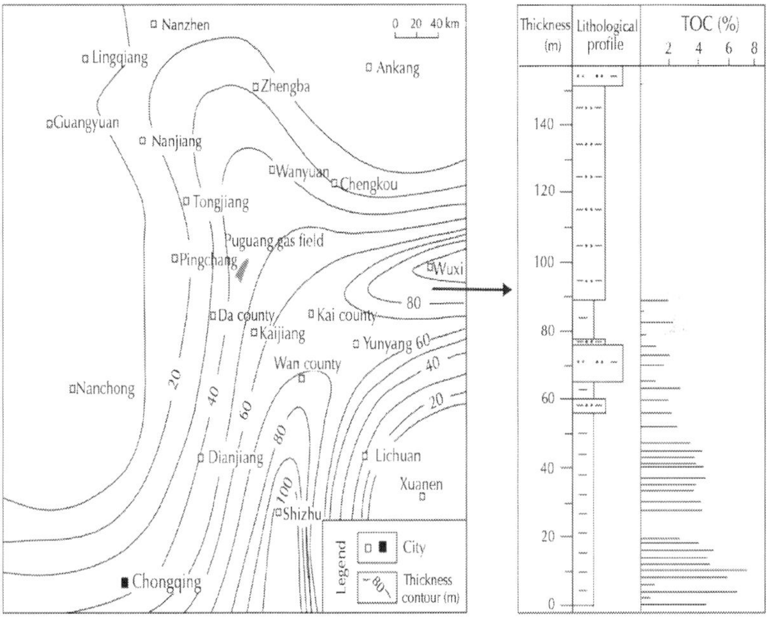

Figure 3: (a) Distribution of hydrocarbon source rocks in the Longmaxi Formation, northeastern Sichuan Basin (unit: m) and (b) variation in TOC contents in a type section of the Longmaxi Formation near Wuxi.

The Longtan Formation consists of around 120 m of carbonate and 200 m of coal-bearing clastic rocks. Because of their low TOC contents, the carbonate rocks are not considered as an effective hydrocarbon source rock. However, coals and dark mudstones with up to 8.2% TOC occurring in eastern Sichuan (e.g. Xuanhan–Daxian region, Fig. 4) show maturity levels around 1.9–3.4% Ro, and the humic nature of their organic matter indicates that these rocks most likely represents the main contributor to the gases in the Feixianguan Formation reservoir. Because of the high thermal maturity level and multiepisodes of tectonic movements in this region, liquid hydrocarbons generated at earlier stages may have undergone thermal cracking to form natural gas and solid bitumen. This suggestion is supported by the widespread occurrence of solid bitumens in many of the reservoirs. As discussed later, the presence of several hydrocarbon source beds also makes it difficult to correlate gases with their source rocks.

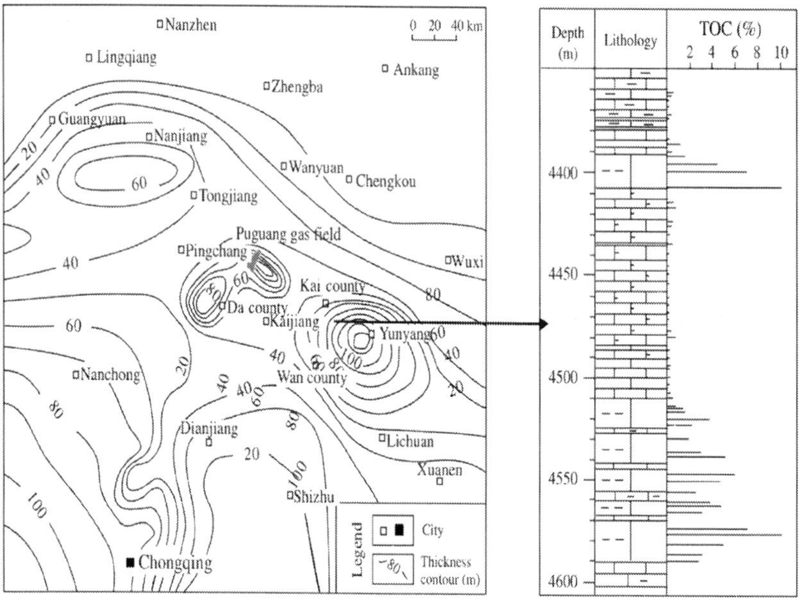

Figure 4: (a) Distribution of hydrocarbon source rocks in the Longtan Formation, northeastern Sichuan Basin (unit: m) and (b) variation in TOC contents in a type section of the Longtan Formation from the Long 4 well.

Gas Reservoirs

During the deposition of the Upper Permian Changxing Formation, the northeastern Sichuan was bounded by the Chengkou–Exi trough in the east and the Kaijiang–Liangping trough in the west (Xu and Zhao, 2003). As a result, reef facies were developed along the two platform edges in the Panglongdong–Yanggudong and Maoba–Puguang regions (Fig. 5a), with the carbonate platform facies in the middle. The Changxing Formation on the Puguang structure consists of sponge bafflestone, sponge framestone, oolitic sparite and bioclastic sparite of the platform edge facies (Ma et al., 2005a). Due to the large-scale marine regression during the late stage of deposition of the Changxing Formation, early reefs or shallow shoals were exposed in topographic highs, leading to formation of thick dolomite rock. Therefore, platform edge reefs and shoals are the favorable sedimentary facies for high-quality oil and gas reservoirs in the Changxing Formation.

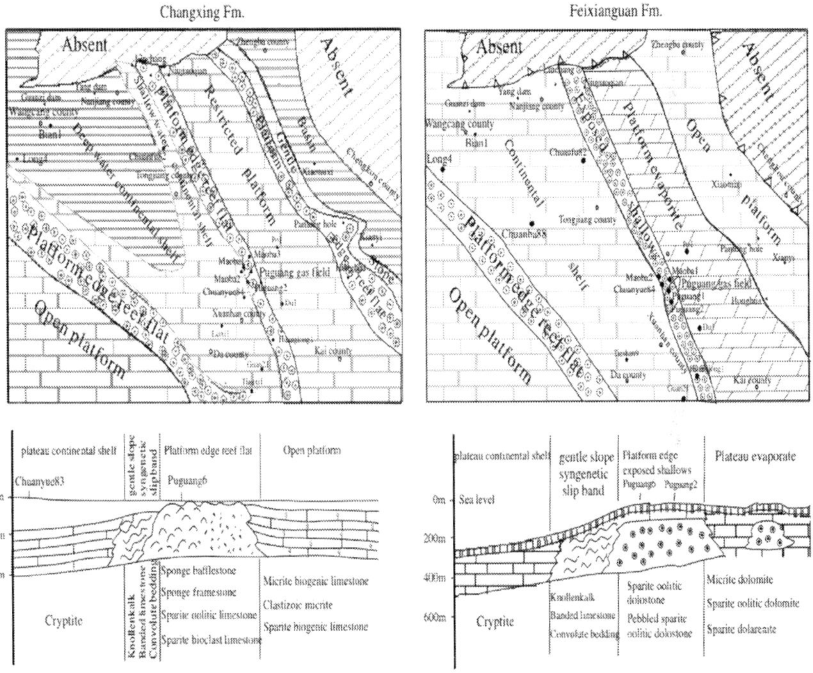

Figure 5: Sedimentary facies map and schematic model for the Changxing (left) and Feixianguan (right) formations in northeastern Sichuan.

During the deposition of the Lower Triassic Feixianguan Formation, the northeastern Sichuan inherited the topographic configuration of the Changxing Formation, with the NW–SE sedimentary facies trends (Fig. 5b). The sediments were largely deposited in an open platform, continental shelf, platform edge shoal (Wang et al., 2002; Yang and Wen, 2002; Wei et al., 2006; Zhu et al., 2006b, Zhu et al., 2006c and Zhu et al., 2006d). The Puguang field lies in the carbonate platform edge, including exposed platform edge shallow shoal, open platform, platformal evaporate facies, ideal sites for high quality petroleum reservoirs. During the deposition of the first to third member of the Feixianguan Formation, the aerial extent of the platform facies expanded gradually, leading to skeletal shoal and oolitic shoal. The third member of the Feixianguan Formation represented the main period of shoal formation in northeastern Sichuan, resulting in sheet-like shoal deposit controlled mainly by tide.

Statistic analyses reveal a strong dependence of reservoir rock porosity with sedimentary facies. The best reservoirs are often found in the shallow oolitic shoal facies deposited in high energy environment. These consist mainly of over 300 m of leached oolitic dolomite and leached sucrosic oolitic dolomite along the slope region. The Puguang 2 well was cored extensively in the Feixianguan Formation from the 4860 to 5200 m interval. Analyses of 700 core samples show that the shallow shoal deposits along the exposed platform edge produce reservoirs with high porosity (10–25%) and high permeability (1–1000 mD).

The prolonged and complex diagenesis of the oolitic shoal deposits has resulted in an improvement of the reservoir quality in the Puguang field. Significant alteration of the carbonates and development of effective secondary porosity resulted from late burial diagenesis involving dedolomitization (and concomitant carbonate leaching) in the pack/wackestone, and the dissolution of anhydrite cement in the grainstone. Late stage recompaction and production of fines in grainstone by the dissolution of dolomite/calcite and anhydrite accompanied this diagenesis. The secondary pores include intercrystaline pores, intracrystaline solution pores, intergranular pores, intragranular solution pores, large solution pores, moldic pores, as well as dissolution clastic textures within the fine-grained dolomite. Analytical data show that the secondary porosity accounts for almost three quarters of the total porosity in the Feixianguan Formation reservoirs. As discussed in later text, thermochemical sulfate reduction (TSR) occurs extensively in the Feixianguan Formation reservoirs, resulting in not only changes in the chemical compositions of the pore fluids, but also the enlargement of corroded pores and holes.

Structure and Trap

Influenced by the Eastern Sichuan Thrust Folding Zone in the east and the Daba Thrust Folding Zone in the north, two types of structural traps were formed in northeastern Sichuan Basin. One is trending in NE–SW direction with steep structure flanks, occurring in the Eastern district (Fig. 1). Another is trending in NW–SE direction, distributed mainly in the western Huaying Mountains and the front ranges of the Daba Mountains in the north. The Matouzhai–Laoyingyan region close to the Daba Mountains generally has compact structures with NEE–SWW strike, and the structures gradually broaden towards the basin direction. In the Huangjinkou and Luojiazhai region, mainly normal faults with NE strike and close folds have been developed. Based on seismic data (Ma et al., 2005a), the lateral distribution of the oolitic reservoirs far exceeds the structural enclosure area, thus forming structure–lithology composite traps.

The Dongyuezhai–Puguang structure belt consists of several NNE–SSW trending faulted anticlines (Fig. 1). The structural top in the southwest is shallower than the northeast, with a steeper flank in the west than in the east. The Puguang field lies in the northern segment of this structural belt, separated from the Dongyuezhai structure by a saddle. Wells drilled on the structural high of the Dongyuezhai structure (Chuanyue 83 and 84 wells) did not encounter high-quality reservoirs and gas shows in the Feixianguan and Changxing formations. However, Puguang 1, 2 and 4 wells drilled in the structural lows in the north flank of the Puguang structure penetrated thick dolomite reservoirs and produced high gas flow (Ma et al., 2005a). The difference appears to be related to a change in depositional environment from platform edge–platform facies to continental shelf facies, as shown clearly from seismic cross sections (Fig. 6). Thus, the Puguang gas trap is a composite structural–lithological trap, controlled by a nose-like structure and sedimentary facies change.

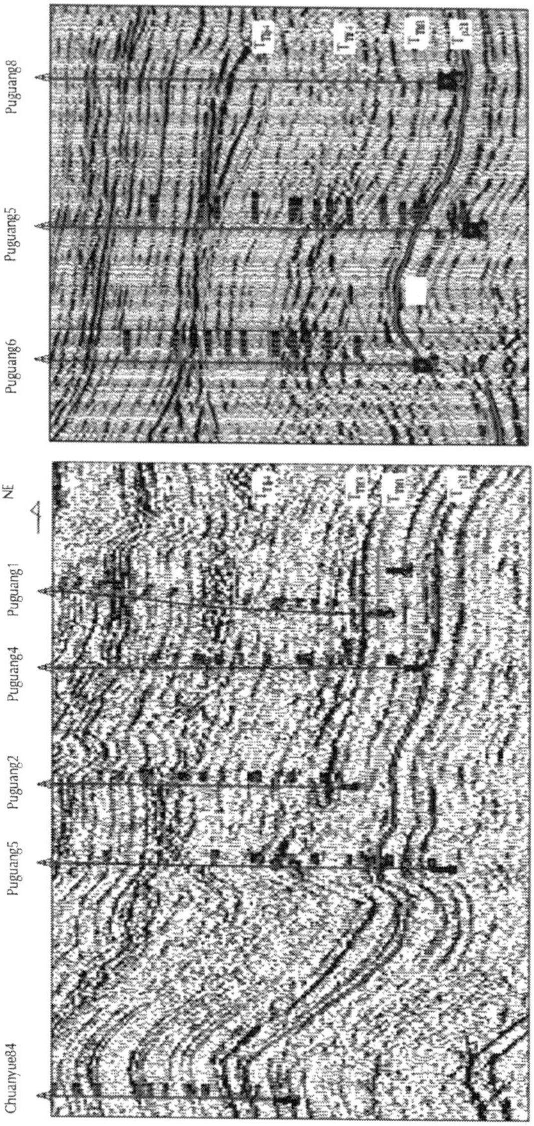

Figure 6: Seismic sections across the Puguang gas field.

As shown in Fig. 1, several other structural–lithological gas pools have been found in the Feixianguan Formation reservoirs on the eastern flank of the Puguang gas field, including the Luojiazhai, Dukouhe and Tieshanpo gas pools. The reservoir depths in these gas pools range from 3500 to 4500 m, much shallower than those in the Puguang field. Gas compositions in these pools are generally similar to those of the Puguang field, indicating similar origin, migration and accumulation history. However, they are much smaller in pool size and gas reserves.

Migration Conduits and Caprocks

There are two important regional caprocks in the northeastern Sichuan. One is the thick mudstone beds in the Upper Triassic Xujiahe Formation and Jurassic strata. They act not only as the direct cap rock for the Triassic–Jurassic nonmarine sequence, but also as the indirect cap rock for the underlying marine carbonate strata. Another is the evaporate beds of the lagoon-tidal flat facies in the Lower and Middle Triassic strata (Fig. 2). These include the fourth member of the Feixianguan Formation, the second and fourth members of the Jialingjiang Formation, and the second member of the Middle Triassic Leikoupo Formation. Due to the tight lithology, strong plasticity, great thickness and regional continuity, these evaporate exhibit excellent sealing capability for the underlying gas accumulations.

The development of the Puguang structure has been closely associated with large reverse faults that cut deeply into the Lower Paleozoic strata. Most of these basement faults were active in the Mesozoic and the early Himalayan orogeny (Wang et al., 1989). Thus, the associated fracture zones were the most likely hydrocarbon migration conduits when the faulting was active, and became effective barriers for fluid flow when the faulting seized. As indicated by seismic reflection data (Fig. 1), these reverse faults terminated within the Feixianguan Formation, thus ensuring the seal integrality of the evaporate beds in the Jiangling and Leikoupo formations (Table 1).

Table 1: Key characteristics of the main hydrocarbon source rocks in eastern Sichuan Basin

Strata	**Lower Cambrian**	**Lower Silurian**	**Upper Permian**
Formation	Jiulaodong	Longmaxi	Longtan
Lithology	Argillaceous rock	Black graptolite-bearing argillutite	Carbargilite and mudstone
Av. TOC (%)	1.67	4.00	5.00
Kerogen type	II	II	II, III
Ro (%)	>3.5	2.0–4.0	1.0–3.4
Thickness (m)	100–350	Av. 300	130–250
Gas generation intensity (bcm/km^2)	7	6–15	10
Peak hydrocarbon generation	Permian	Middle Triassic	Middle Jurassic

Reservoir Fluid Chemistry and Origin

Natural gases produced from the Puguang gas field are typically dry, with around 80% methane, less than 0.05% C_{2+} alkanes, and $C_1/(C_1–C_5)$ ratios greater than 0.995 (Huang, 1984). These gases are rich in H_2S (14–17%) and CO_2 (3–10.5%), with similar chemical compositions for gases produced from the Feixianguan and Changxing formations (Table 2). The ^{13}C values of methane and ethane in these gases are within the range of −30.0‰ to −31.0‰ and −25.2‰ to −29.1‰ respectively, consistent with a dominantly coaly source (Zhu et al., 2006a, Zhu et al., 2006b, Zhu et al., 2006c, Zhu et al., 2006d and Zhu et al., 2006e), perhaps with additional contribution from high-temperature oil cracking (Fig. 7). The presence of extensive pyrobitumens in the Feixianguan Formation reservoirs provides clear evidence for an early oil charge and subsequently thermal cracking under higher temperatures. Based on the 1-D basin modeling results in the study area (Fig. 8), peak hydrocarbon generation for the Cambrian, Lower Silurian and Upper Permian source rocks occurred in Permian, Middle Triassic, and Middle-Late Jurassic, respectively. Hydrocarbon fluid inclusion data (Table 3) indicate three separate charge times to the Feixianguan Formation reservoirs, corresponding to the late episode of the Indo-Sinian orogeny (Fig. 8). Thus, it is possible that the early oil charge was derived from type II hydrocarbon source rocks in the Lower Silurian Longmaxi Formation, whereas the gases came from all of the three possible sources, perhaps with significant proportions from the coal measures in the Upper Permian Longtan Formation as well as the in-reservoir oil cracking associated with TSR.

Table 2: Chemical and stable carbon isotope compositions of the gases produced from the Puguang gas field

Well	**Formation**	**Depth (m)**	**Composition** (‰)								**(C_1/ C_{1-4})**	**$\delta^{13}C$** (‰)	
			CH_4	C_2H_6	C_3H_8	H_2S	CO_2	N_2	H_2	He		CH_4	C_2H_6
Puguang1	T_1f	5610	77.5	0.02	0.000	12.7	9.10	0.62	0.04	0.01	0.9999		
Puguang 2	T_1f	5062	75.6	0.11	0.000	15.8	7.96	0.44	0.03	0.01	0.999	–31.0	–28.8
	T_1f	4958	74.3	0.22	0.000	17.2	7.9	0.42	0.01	0.01	0.997	–30.5	–29.1
	T_1f	4801	76.1	0.02	0.000	15.5	7.71	0.44	0.01	0.01	0.9999	–30.9	–28.5
	P_2ch	5315	74.2	0.02	0.000	16.0	9.46	0.45	0.00	0.01	0.9999	–30.6	–25.2
	P_2ch	5259	75.0	0.33	0.000	15.4	8.73	0.47	0.02	0.01	0.996	–30.1	–27.7
Puguang 4	T_1f	5770	73.8	0.03	0.000	17.1	8.47	0.59	0.00	0.02	0.9999		
Puguang 6	T_1f	5044	74.7	0.03	0.000	14.1	10.5	0.65	0.06	0.01	0.9999		
	P_2ch	5240	75.9	0.05	0.000	14.7	8.74	0.49	0.06	0.01	0.999		

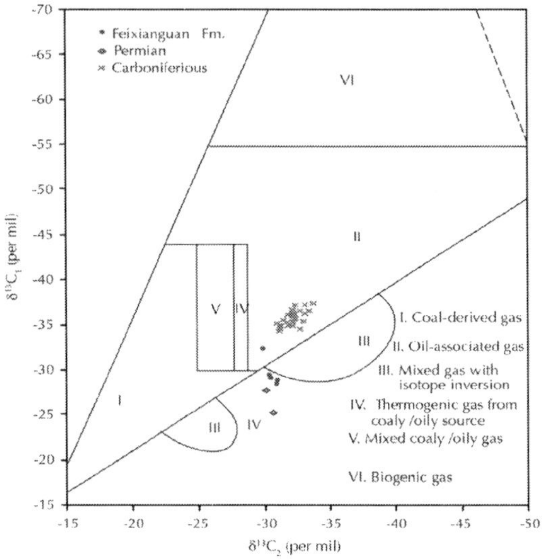

Figure 7: Genetic classification of natural gases in northeastern Sichuan, based on the scheme proposed by Dai (1985).

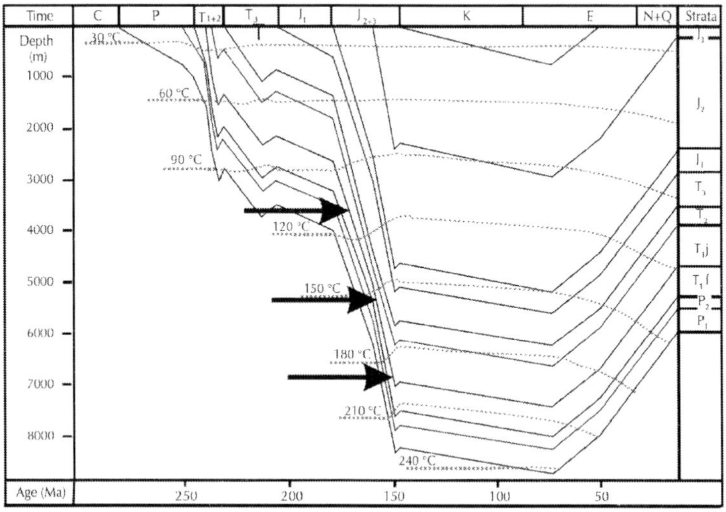

Figure 8: Burial and thermal history of the major stratigraphic intervals at the location of Puguang 2 well. Critic moments for oil and gas charge were determined using the homogenization temperatures of aqueous in-

clusions associated hydrocarbon fluid inclusions. The model was built using Petromodel software. Compaction and denudation were based on wireline log data, and thermal maturity model calibrated using measured bottom hole temperatures and vitrinite reflectance data.

Table 3: A summary of aqueous fluid inclusion data for samples collected from the Puguang 2 well

Host mineral	Type	Color and shape	Homogenization temperature (°C)	Peak value (°C)	Freezing point (°C)	Salinity (%)	Filling time
Calcite in early fracture and granular calcite	Liquid and solid	Abnormally light blue fluorescence	100–120	110	−8.9 to −9.4	12.86–13.30	Late Indo-Sinian to Early Yanshanian
Late calcite intergrowth, calcite vein and dissolved quartz	Mainly Gas/liquid phases with some liquid phase only	Rectangular rotundity, ellipse and long ellipse	140–160	150	−3.3 to −7.6	5.32–9.84	Middle-late Yanshanian
Calcite vein and dissolved quartz	Gas-liquid phases, or gas mostly	Long ellipse, rotundity, square and abnormity	170–210	190	−7.5–8.0	11.10–11.70	Himalayan

There have been many discussions on the potential sources of H_2S in petroleum reservoirs (e.g. Orr, 1974;Machel, 2001; Worden et al., 1996; Krouse et al., 1988; Heydari, 1997; Goldhaber and Orr, 1995; Dai, 1985; Rooney, 1995; Zhang et al., 2005). Because bacterial sulfate reduction and thermal decomposition of sulfur-rich sedimentary organic matter generally produce only 3–5% H_2S, these mechanisms have been excluded as the dominant cause for the sour gases reservoired in the Feixianguan and Changxing formations in northeastern Sichuan (Cai et al., 2004; Li et al., 2005; Zhu et al., 2005a and Zhu et al., 2005b). Due to the periodic change in sea level and the blockage of seawater by oolitic shoal and bar, a few thin anhydrite-rich beds were deposited within the main reservoir intervals in the study area, with a cumulative thickness about 5–10 m. As shown in Fig. 9, the main reservoirs in the Puguang area were buried at least 1000–2000 m deeper in the Mesozoic period than their current depth, and the reservoir temperatures reached at least 120 and 180 °C in Middle Jurassic and Late Cretaceous, respectively. Thus, favorable conditions were present for TSR to occur in these reservoirs.

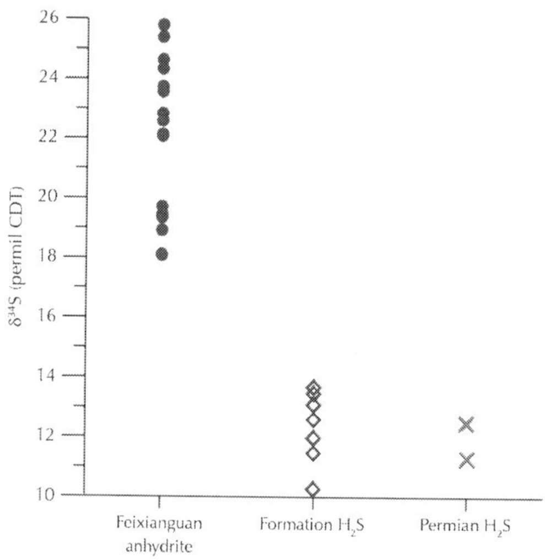

Figure 9: Variation in the $\delta^{34}S$ values for the anhydrite and produced H_2S in the Feixianguan and Changxing formations in northeastern Sichuan.

In order to assess the origin of H_2S as well as the cause of petroleum alteration, H_2S concentrations and the $\delta^{34}S$ values have been collected and measured in gas samples from the reservoirs. Anhydrite, pyrite and elemental sulfur $\delta^{34}S$ values have been measured for comparison. Where H_2S concentrations are greater than 10% its $\delta^{34}S$ values range from +10.28 to +13.71‰ CDT.

These are lower than the range of anhydrite $\delta^{34}S$ values found within the Feixianguan Formation (+18.09 to +25.08‰ CDT, Fig. 9), suggesting an isotopic fractionation around 6.5–14‰ (averaged at 10‰). This fractionation is larger than that observed from the Upper Devonian sour gas reservoirs in Alberta, Canada (Krouse et al., 1988). Under isotopic equilibrium conditions, sulfur isotopes split unevenly between the sulfate and the sulfide phases with more ^{34}S going into the oxidized form. Laboratory experiments have shown that during the transformation of sulfates to sulfides (e.g. during TSR), the sulfur isotope fractionation decreases rapidly with increasing temperature. The fractionation factor is around 15‰ at 150 °C and 10‰ at 200 °C (Machel, 2001). Thus, under isotopic equilibrium, the reduced sulfur phase should be about 15‰ lower than the oxidized sulfur phase at 150 °C. However, it has been found and reported from many basins around the world that, under natural conditions in sedimentary basins, TSR leads to identical $\delta^{34}S$ values for the initial sulfate and resulting reduced sulfur compounds (e.g. Worden et al., 1996; Machel, 2001; Cai et al., 2004). Thus, the rate of TSR is controlled not by the chemical redox step but by either a dissolution or a transport step. Therefore, it remains to be determined if the 6.5–14‰ shift in the $\delta^{34}S$ values of the H_2S and anhydrite in the Puguang gas field resulted from the cumulative effect of chemical reactions ranging from the earlier BSR to later TSR over a prolonged time span of the given reservoirs. For this reason, petrographic study is currently being conducted in order to understand which processes have been involved. As ethane, propane and even methane may have been involved in TSR reactions, one needs to be extremely cautious in the use of their $\delta^{13}C$ values in gas source correlation.

Charge History

Our preliminary investigation indicates three possible source rocks for the natural gases in the Puguang gas field, including the coal measures in the Upper Permian Longtan Formation and the highly mature source rocks in the Lower Cambrian. Liquid hydrocarbons derived from the Lower Silurian Longmaxi Formation may have also been thermally cracked in local reservoirs to produce additional gases. As discussed in earlier section, source rocks in the Longmaxi Formation reached peak oil generation in Middle Triassic (Indo-Sinian orogeny) and peak gas generation in the mid and late episodes of the Yanshan orogeny (J–K). Unexpelled residual liquid hydrocarbons in these source rocks would have been thermally cracked to form gas by the Late Cretaceous. Hydrocarbon fluid inclusion data (Li et al., 2005; Zhu et al., 2006a) show that the reservoir bitumens most likely represent an early oil charge in Middle Triassic that were subsequently transformed into gas and pyrobitumen that are common in gas reservoirs. In contrast, source rocks in the Longtan Formation reached their peak stage of hydrocarbon generation in Middle and Late Jurassic. Thus, large-scale hydrocarbon expulsion, migration and accumulation in the study area occurred most likely during the Yanshan orogeny (J–K), slightly later than that for structural formation in the Triassic time.

Back-striping results (Fig. 10) suggest that the original configuration of the Dongyuezai–Puguang structure belt was characterized by a northeastern high and a southwestern low. As the Lower Silurian source rocks entered oil window toward the end of the Triassic, an oil accumulation was formed along the structural high at the present Puguang–Tieshanpo area. Some of the hydrocarbons generated from the Lower Cambrian may have also been trapped. During the Yanshan (J–K) and Himalayan orogenies (E–N), the peripheral trust folding gradually advanced toward the basin, and the increase in sedimentary loading in the north resulted in a subsiding foreland basin. As a result, the dipping of the Dongyuezai–Puguang structure belt was reversed, and the early oil/gas reservoirs were subjected to undergo further thermal maturation and TSR reactions. At the same time, coal measures in the Upper Permian Longtan Formation also entered peak hydrocarbon generation and expelled significant volumes of gas. These gases have been largely trapped in the Changxing and Feixianguan formations because of the thick evaporate beds intercalated with and/or above the oolitic reservoir rocks. During the late episode of the Himalayan orogeny (N), the region was uplifted, leading to structural adjustment and denudation. Along with the structural deformation and reduced loading, a composite structural–lithological gas pool emerged in the present structure location.

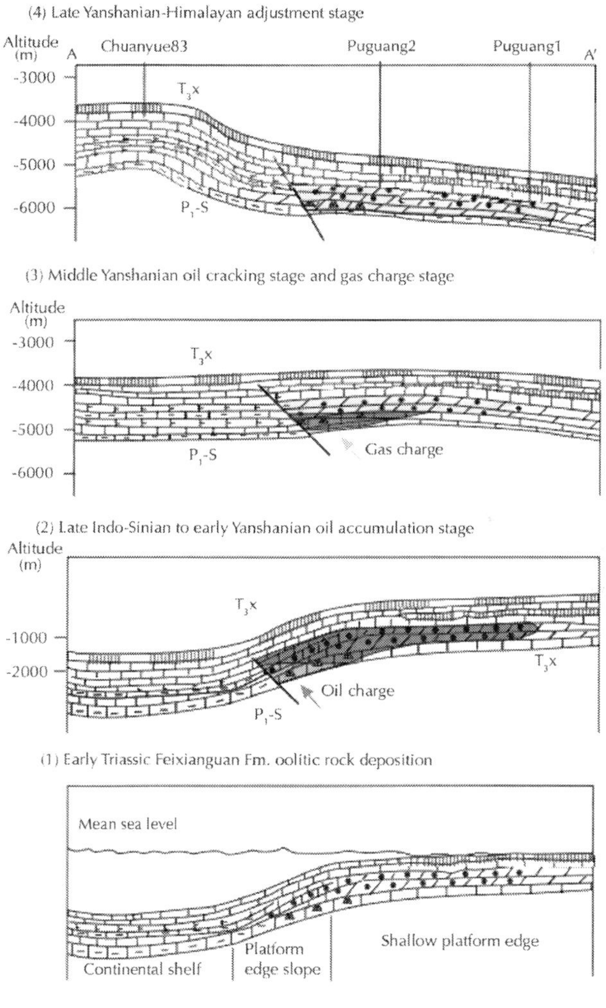

Figure 10: Schematic models showing the evolution of the structure and reservoir fluids in the Feixianguan Formation, Puguang gas field.

Based on the reconstructed structural configurations over geological time, the evolution of the Puguang gas field can be divided into three stages:

Stage 1: Oil charge in the early episode of the Yanshan orogeny. In this period, the Lower Silurian source rocks entered oil window, and gentle anticlines were formed in the basin. Organic acids expelled from the hydrocarbon source rocks partially dissolved the oolitic rocks, forming early burial solution pores in vantage position and improving reservoir porosity. The combination of the anticline structure with excellent oolitic reservoirs provided excellent habitat for liquid hydrocarbons.

Stage 2: In-reservoir thermal maturation and gas injection in the middle and late episodes of the Yanshan orogeny. Hydrocarbon source rock entered into gas window for persistent subsidence and deep burial. The early charged oil in reservoir underwent thermal cracking to form pyrobitumens and gas. The accompanied TSR reactions produced significant quantities of H_2S, and led to modified pore structures and perhaps increased porosity in reservoir. Coal measures in the Upper Permian Longtan Formation released large volume of gas into the overlying reservoirs in the Changxing and Feixianguan formations. These hydrocarbon fluids migrated toward structural high to form a dominantly gas pool.

Stage 3: Readjustment in gas trap configuration during the Himalayan orogeny. Uplift and denudation reduced the burial depth and temperature of gas reservoir, and eventually terminated the TSR reactions. The change in the regional tectonic stress orientation over time led to the structural reconfigurations and reservoir fluid readjustment.

SUMMARY

Key geological elements that have contributed to the formation of the giant Puguang sour gas field in northeastern Sichuan Basin include:

- Excellent hydrocarbon source rocks in the coal measures in the Upper Permian Longtan Formation, the graptolite-bearing black shales in the Lower Silurian Longmaxi Formation, and the dark mudstones in the Lower Cambrian Jiulaodong Formation.
- High quality reservoir rocks in the Lower Triassic Feixianguan Formation and Upper Permian Changxing Formation, controlled primarily by the oolitic shoal facies developed on the evaporitic carbonate platform edge, and further improved by secondary diagenetic processes, such as the late burial dolomitization, deep burial organic acid corrosion, and TSR.
- Evaporate beds in the Lower and Middle Triassic strata acting as the direct top seal for the Upper Permian–Lower Triassic gas accumulations, and the thick Upper Triassic and Jurassic mudstones as an indirect cap rock for the marine carbonate strata.
- The deep-seated basement faults as conduits for vertical hydrocarbon migration during faulting, becoming effective flow barriers when the faulting ceased.
- The development of a series of anticlinal structures coupled with the high quality oolitic shoal reservoirs provided the necessary conditions for the formation of composite structural–lithological traps.
- The deep burial following the initial oil charge in reservoirs containing abundant sulfate minerals resulted in extensive TSR reactions to form large quantities of thermogenic methane, H_2S and reservoir bitumens.
- TSR reactions, new gas injection and structural reconfiguration transformed an old oil pool to a large sour gas pool.

Looking back at the discovery process of the northeastern Sichuan, it becomes clear that knowing the complex interplays of various petroleum system elements holds the key for the exploration success. With improved geological understanding, creative exploration strategies and innovative engineering technology, we believed that the giant Puguang gas field represented only one of the first major discoveries to be made in the Sichuan Basin. This idea has been proven through the recent discovery of the Longgang giant gasfield in northeastern Sichuan Basin in March 2006 by PetroChina Company.

ACKNOWLEDGMENTS

The authors extend their gratitude to numerous staff at SINOPEC Southern E&P Company for their years of hard work and technical support. We thank MPG reviewers Hairuo Qing (University of Regina) and Volker Dieckmann (Shell International Company) for their helpful comments.

REFERENCES

1. Cai, C.F., Xie, Z.Y., Worden, R.H., Hu, G.Y., Wang, L.S., He, H., 2004. Methane-dominated thermochemical sulphate reduction in the Triassic Feixianguan formation East Sichuan Basin, China: towards prediction of fatal H2S concentrations. Marine and Petroleum Geology 21, 1265–1279.
2. Dai, J.X., 1985. Distribution, classification and origin of natural gas with hydrocarbon sulphide in China. Acta Sedimentologica Sinica 3 (4), 109–120 (in Chinese).
3. Dai, J.X., Wang, T.D., Dai, H.M., 2000. Gas sources of large-scale carbonate gasfield. Marine Petroleum Geology 5 (1/2), 143–144 (in Chinese).
4. Dai, J.X., Xia, X.Y., Wei, Y.Z., Tao, S.Z., 2001. Carbon isotope characteristics of natural gas in the Sichuan Basin, China. Experimental Petroleum Geology 23 (2), 115–121 (in Chinese).
5. Dai, J.X., Qin, S.F., Tao, S.Z., Zhu, G.Y., Mi, J.K., 2005. Developing trends in natural gas industry and new advances in natural gas geological theories in China. Natural Gas Geoscience 16 (2), 127–142 (in Chinese).

6. Goldhaber, M.B., Orr, W.L., 1995. Kinetic controls on thermochemical sulfate reduction as a source of sedimentary H2S. In: Vairavamurthy, M.A., Schoonen, M.A.A. (Eds.), Geochemical Transformations of Sedimentary Sulfur, ACS Symposium Series, vol. 612, pp. 412–425.
7. Heydari, E., 1997. The role of burial diagenesis in hydrocarbon destruction and H2S accumulation, Upper Jurassic smackover formation, Black Creek Field, Mississippi. American Association of Petroleum Geologists Bulletin 81, 26–45.
8. Hu, G.C., Xie, Y.X., 1997. Carboniferous Reservoirs in the High and Steep Structure Eastern Sichuan Basin. Petroleum Industry Press, Beijing (in Chinese).
9. Huang, J.Q., 1954. On Major Tectonic Units of China. Geological Press, Beijing, 161pp. (in Chinese).
10. Huang, J.Z., 1984. Geochemical characteristics of natural gases in the Sichuan Basin. Geochimica 4, 307–321 (in Chinese).
11. Jin, Z.J., 2005. Particularity of petroleum exploration on marine carbonate strata in Chinese sedimentary basins. Earth Science Frontiers 12 (3), 15–22 (in Chinese).
12. Krouse, H.R., Viau, C.A., Eliuk, L.S., Ueda, A., Halas, S., 1988. Chemical and isotopic evidence of thermochemical sulphate reduction by light hydrocarbon gases in deep carbonate reservoirs. Nature 333, 415–419.
13. Li, J., Xie, Z.Y., Dai, J.X., Zhang, S.C., Zhu, G.Y., Liu, Z.L., 2005. Geochemistry and origin of sour gas accumulations in the northeastern Sichuan Basin, SW China. Organic Geochemistry 36 (12), 1703–1716.
14. Li, J.C., Ma, Y.S., Zhang, D.J., 1998. Some important scientific problems on exploration in marine formations of China. Petroleum Exploration and Development 25 (5), 1–2 (in Chinese).
15. Liu, S.G., Xu, G.S., Liang, W., Wang, Y.C., Zhang, G.X., Long, C.J., Wang, Z.Y., 1997. A study on gas system of Carboniferous

gas pools in eastern Sichuan Basin. Acta Petrolei Sinica 18 (3), 13–22 (in Chinese).

16. Ma, Y.S., Cai, X.Y., Li, G.X., 2005a. Basic characteristics and zoning of the Puguang gas field in the Sichuan Basin. Acta Geological Sinica 79 (6), 858–865 (in Chinese).
17. Ma, Y.S., Guo, X.S., Guo, T.L., Huang, R., Cai, X.Y., Li, G.X., 2005b.
18. Discovery of the giant Puguang gas field in the Sichuan Basin and implications for future exploration. Geological Review 51 (4), 477–480 (in Chinese).
19. Ma, Y.S., Mu, C.L., Guo, X.S., Tan, Q.Y., Yu, Q., 2006. Characteristic and framework of the Sedimentation of the Changxing formation in the northeastern Sichuan Basin. Geological Review 52 (1), 25–29 (in Chinese).
20. Machel, H.G., 2001. Bacterial and thermochemical sulfate reduction in diagenetic settings—old and new insights. Sedimentary Geology 140, 143–175.
21. Orr, W.L., 1974. Changes in sulfur content and isotopic ratios of sulfur during petroleum maturation—study of the big horn basin Paleozoic oils. AAPG Bulletin 50, 2295–2318.
22. Ran, L.H., Chen, G.S., Xu, R.F., 2005. Discovery and exploration of Luojiazhai Gas Field, Sichuan Basin. Marine Origin Petroleum Geology 10, 43–47.
23. Rooney, M.A., 1995. Carbon isotopic ratios of light hydrocarbons as indicators of thermochemical sulfate reduction. In: Grimalt, J.O. (Ed.), Organic Geochemistry: Applications to Energy, Climate, Environment and Human History. San Sebastian, pp. 523–525.
24. Wang, B.J., 1989. The Sichuan Hydrocarbon Province. Volume 10 of Petroleum Geology of China. Petroleum Industry Press, Beijing, p. 516 (in Chinese).
25. Wang, B.J., Bao, C., Lou, Z., Guo, Z., 1989. Formation and development of the Sichuan Basin. In: Zhu, X. (Ed.), Chinese Sedimentary Basin. Elsevier, Amsterdam, pp. 147–164.

26. Wang, M., Zhang, G.B., Zhu, Z.G., 2005. Assessment of Oil and Gas Resources of China-2005. The Office of National Oil & Gas Resource Assessment, Beijing (in Chinese).
27. Wang, Y.Y., Liu, H.Y., Wen, Y.C., Yang, Y., Zhang, J., 2002. Distribution law, exploration method and prospect of the oolitic beach reservoirs in the Feixianguan formation in northeast Sichuan Basin. Natural Gas Industry 22 (Suppl.), 14–18 (in Chinese).
28. Wei, G.Q., Chen, G.S., Yang, W., Yang, Y., Hu, M.Y., Zhang, L., Wu, S.X., Jin, H., Shen, J.H., 2006. Sedimentary system and evolution of platformal trough of the Lower Triassic Feixianguan Formation in northern Sichuan Basin. Acta Sedimentologica Sinica 22 (2), 254–260 (in Chinese).
29. Worden, R.H., Smalley, P.C., Oxtoby, N.H., 1996. The effects of thermochemical sulfate reduction upon formation water salinity and oxygen isotopes in carbonate reservoirs. Geochimica et Cosmochimica Acta 60, 3925–3931.
30. Xu, G.S., Zhao, Y.H., 2003. Analysis on the formation mechanisms for the Carboniferous gas reservoirs in the Kaijiang paleohigh region of east Sichuan area. Experimental Petroleum Geology 25 (2), 158–163 (in Chinese).
31. Yang, Y., Wen, Y.C., 2002. Control of the Kaijiang–Liangping trough growth on the T1f oolitic rocks in northeastern Sichuan Basin. Natural Gas Industry 22 (Suppl.), 30–32 (in Chinese).
32. Zhai, G.M., 1992. Petroleum Geology of China. Petroleum Industry Press, Beijing (in Chinese).
33. Zhang, S.C., Liang, D.G., Zhang, D.J., 2002. Evaluation criteria for Paleozoic effective hydrocarbon source rocks. Petroleum Exploration and Development 29 (2), 8–12.
34. Zhang, S.C., Zhu, G.Y., Liang, Y.B., Dai, J., Liang, H., Li, M., 2005. Geochemical characteristics of the Zhaolanzhuang sour gas accumulation and thermochemical sulfate reduction in the Jinxian Sag of Bohai Bay Basin. Organic Geochemistry 36 (11), 1717–1730.

35. Zhang, S.C., Zhu, G.Y., Liang, Y.B., 2006. Sour gas origin and highquality gas reservoirs in the Puguang giant gas field in Sichuan. Geological Review 52 (2), 46–50 (in Chinese).
36. Zhao, W.Z., Luo, P., Chen, G.S., 2005. Dolomitization and reservoir quality in the Lower Triassic Feixiangan Formation, NE Sichuan Basin, China: significance for future gas exploration. Journal of Petroleum Geology 28 (1), 83–100.
37. Zhu, G.Y., Zhang, S.C., Li, J., Jin, Q., 2004. Formation and distribution of hydrogen sulfide bearing gas in China. Petroleum Exploration and Development 31 (4), 18–21 (in Chinese).
38. Zhu, G.Y., Zhang, S.C., Liang, Y.B., 2005a. Isotopic evidence of TSR origin for H2S-rich gases in the Feixianguan Formation, northeastern Sichuan Basin. Science in China 48 (11), 1960–1971 (in Chinese).
39. Zhu, G.Y., Zhang, S.C., Liang, Y.B., 2005b. Relationship between palaeoenvironment and H2S distribution in the Feixianguan Formation, NE Sichuan Province. Petroleum and Exploration and Development 32 (4), 34–39 (in Chinese).
40. Zhu, G.Y., Zhang, S.C., Liang, Y.B., Dai, J.X., 2006a. Stable sulfur isotopic composition of hydrogen sulfide and its genesis in Sichuan Basin. Geochimica 35 (4), 412–422 (in Chinese).
41. Zhu, G.Y., Zhang, S.C., Liang, Y.B., 2006b. Formation mechanism and distribution pattern of the high-quality marine reservoirs in deeper Sichuan Basin. Petroleum Exploration and Development 33 (2), 161–166 (in Chinese).
42. Zhu, G.Y., Zhang, S.C., Ma, Y.S., Liang, Y.B., Guo, T.L., Zhou, G.Y., 2006c. Effectiveness of thermochemical sulfate reduction on oil and gas industry—a. H2S formation accelerates secondary porosity in reservoirs. Earth Science Frontiers 13 (3), 142–149 (in Chinese).
43. Zhu, G.Y., Zhang, S.C., Liang, Y.B., Ma, Y.S., Dai, J.X., Zhou, G.Y., 2006d. Dissolution and alteration of the deep carbonate reservoirs by TSR an important type of deep-buried

high-quality carbonate reservoirs in the Sichuan Basin. Acta Petroleum Sinica 22 (8), 2182–2194 (in Chinese).

44. Zhu, G.Y., Zhang, S.C., Liang, Y.B., Ma, Y.S., Zhou, G.Y., Dai, J.X., 2006e. Characteristics of sour gas reservoirs in northeastern Sichuan Basin and TSR effects. Acta Sedimentologica Sinica 24 (2), 79–85 (in Chinese).

Citations

CHAPTER 1

Nanjun Lai, Xiaoping Qin, Zhongbin Ye, Qin Peng, Yan Zhang, and Zheng Ming, "Synthesis and Evaluation of a Water-Soluble Hyperbranched Polymer as Enhanced Oil Recovery Chemical," Journal of Chemistry, vol. 2013, Article ID 824785, 11 pages, 2013. doi:10.1155/2013/824785.

CHAPTER 2

Zhongbin Ye, Xiaoping Qin, Nanjun Lai, Qin Peng, Xi Li, and Cuixia Li, "Synthesis and Performance of an Acrylamide Copolymer Containing Nano-SiO2 as Enhanced Oil Recovery Chemical," Journal of Chemistry, vol. 2013, Article ID 437309, 10 pages, 2013. doi:10.1155/2013/437309.

CHAPTER 3

Bieby Voijant Tangahu, Siti Rozaimah Sheikh Abdullah, Hassan Basri, Mushrifah Idris, Nurina Anuar, and Muhammad Mukhlisin, "A Review on Heavy Metals (As, Pb, and Hg) Uptake by Plants through Phytoremediation," International Journal of Chemical Engineering, vol. 2011, Article ID 939161, 31 pages, 2011. doi:10.1155/2011/939161.

CHAPTER 4

Paul A Siratovich, Michael J Heap, Marlène C Villenueve, James W Cole, and Thierry Reuschlé, Physical Property Relationships of the Rotokawa Andesite, a Significant Geothermal Reservoir Rock in the Taupo Volcanic Zone, New Zealand, doi:10.1186/s40517-014-0010-4.

CHAPTER 5

Albina Mukhametshina and Elena Martynova, "Electromagnetic Heating of Heavy Oil and Bitumen: A Review of Experimental Studies and Field Applications," Journal of Petroleum Engineering, vol. 2013, Article ID 476519, 7 pages, 2013. doi:10.1155/2013/476519.

CHAPTER 6

Amin Beiranvand Pour and Mazlan Hashim, ASTER, ALI and Hyperion sensors data for lithological mapping and ore minerals exploration, doi:10.1186/2193-1801-3-130.

CHAPTER 7

Yongsheng Ma, Shuichang Zhang, Tonglou Guo, Guangyou Zhu, Xunyu Cai, Maowen Li, Petroleum geology of the Puguang sour gas field in the Sichuan Basin, SW China, Marine and Petroleum Geology, Volume 25, Issues 4–5, April–May 2008, Pages 357-370, ISSN 0264-8172, http://dx.doi.org/10.1016/j.marpetgeo.2008.01.010.

Index